PROGRAMMABLE HEURISTICS AND BIG IDEAS

© 2017 NATHAN COPPEDGE

INTRODUCTION-------------P. 9

USING CAT DEDUCTIONS —-P. 13
INFORMAL HEURISTICS——---P. 14

PROGRAM. HEURISTICS——— P. 15
SHORT & MISC------P. 17
CATEGORICAL DEDUCTION —— p. 23
FORMULA FOR PARADOXES———p. 28
PSYCHIC PREDICTION———————p. 30
FORMULA FOR VISUAL CREATIVITY——p. 40
FORMULA FOR SOULS-----p. 42
FORMULA FOR QUESTIONS----P. 45
FORMULA FOR PERFECT ANSWERS——— p. 47
WIZARD LOGIC———————————p. 49
INCOHERENT DEDUCTION——————p. 59
THEORY OF EVERYTHING——————-p. 61

GODLIKE SYSTEMS———-P. 69
WEIGHING OF SOULS———p. 69
NIRVANA——————-p. 73
CREATING MATTER——p. 77
MAGIC TORNADO———p. 78
LOGIC OF POETRY——p. 79
INVENTION——-p. 81
DIVINE FUN——-p. 90
CRAFTING———p. 91
CLARITY———-p. 95
5TH BARRIER——p. 96
ANTIGRAY——-p. 99
MAGIC AC———p. 100

MASTER SYSTEMS————P. 101
IMPROV————p. 102
EVOLVE————p. 106
SAVEGAME————p. 111
FLASHFORWARD————p. 112

MASTER LOGICS
NO DAMAGE————p. 113
ENERGY————p. 114
IMPROVED SYLLOGISMS————p. 116
INFERENCE -------p. 122
X-ICAL DEDUCTION————p. 123
MASTER DEDUCTION---p. 124
HARD LOGIC————p. 126
OLD LOGIC————p. 128

SECONDARY SYSTEMS, P. 129
COINCIDENTAL LOGIC————p. 129
MAGIC SOLVER————p. 130
INTUITION————p. 131
PRIVATE LANGUAGE————p. 132
1ST & 2ND WORLD PROBLEMS————p. 133
INVENTION————p. 134
FANTASY————p. 135
THE GOOD————p. 136
VARIABLE ANALYSIS————p. 139
SEARCH ALGORITHM————p. 140
LACK OF INFORMATION————p. 141

NEGATIVITY STUDIES———P. 143
EPITAPHS————p. 144
UNDESIRABILITY———p. 145
BROKEN REASON————p. 146
EVIL————p. 147

PROGRAMMABLE HEURISTICS AND BIG IDEAS

DEATH———p. 148
FOOLS———p. 149
ANIMALISM———p. 150
GHOSTS———p. 151
FALSEHOODS———p. 152
UNFAIRNESS———p. 153
GUILT———p. 154
ENTROPY———p. 155
NONSENSE———p. 156
FORMAL ZERO———p. 157
FRUSTRATION———p. 159
STUPIDITY———p. 160
MEANING WITHOUT MEANING———p. 163
LIES THAT AREN'T LIES———p. 164
RELATIVISM———p. 166
PROBLEMS———p. 167
WORSE THAN PAIN———p. 168
TRAGEDY———p. 169
DESTRUCTION———p. 170
ANGER———p. 171
SIN———p. 172
KILLING NAZIS———p. 173

OTHER SYSTEMS———P. 175
QUANTUM PERCEPTION———p. 175
DIALECTICAL INFERENCE———p. 176
SUBJECTIVITY———p. 178
ABSOLUTENESS———p. 179
EXCEPTIONS———p. 182
MYSTERY———p. 184
CHILD LOGIC———p. 185
TOO LOGICAL LOGIC———p. 186
DESIGN———p. 187
PHYSICAL PARADIGMS———p. 191
CALCULUS OF FREE WILL --p. 192

SECRET LAWS————P. 195
NOTES——-P. 224

PROGRAM. HEURISTICS P2—- P. 225
ADVANCED PROGRAM. HEUR.— P. 243
ENGLISH HEURISTICS———— P. 246
INFINITE DIMENSIONAL HEUR— P. 250
HEURISTIC SCHMOOZLE ———- P. 252
AGING, HEALTH, AND LONGVITY P. 254
HEURISTICS FOR HAPPINESS———P. 259
FREE WILL HEURISTICS———P. 261
REASONING HEURISTICS——— P. 263

REFERENCES ----P. 265

BIGGEST IDEAS —————P. 269
BIG IDEAS —————--— P. 291

ANY MORE ZERO-TO-ONE IDEAS?—-P. 332

RECOMMENDED READING ----P. 343

BIO-----------------------------P. 346

PROGRAMMABLE HEURISTICS

AND BIG IDEAS

—By Nathan Coppedge

Nathan Coppedge

PROGRAMMABLE HEURISTICS AND BIG IDEAS

INTRODUCTION

Nathan Coppedge

PROGRAMMABLE HEURISTICS AND BIG IDEAS

• INTRODUCTION •

Alan Hájek, who earned his PhD. in philosophy at Princeton in 1993, argues that "philosophy has a variety of heuristics" (1). And goes on to say that "such heuristics can enhance one's ability to make creative contributions to philosophy." (1).

As recently as 1972, Karl Popper did not even consider philosophical objectivity (in other words, *progress*) possible. He writes:

"From a rational point of view, we should not 'rely' on any theory, for no theory has been shown to be true, or can be shown to be true." (Popper 21).

This is a strange absoluteness to hold, which must be self-contradictory.

Typology, which is perhaps the closest to realizing coherence outside of mathematics, emerged as recently as the 1870s (145 years ago) with Charles Peirce. ("This theory of meaning ('speculative grammar') was to provide foundations for his writings in logic" OCP).

Nicholas Rescher's book (1973), dates work on coherence somewhat more recently.

However, not all heuristics involve coherence. I have made an effort to include anything that might be helpful to my readers.

Nathan Coppedge

PROGRAMMABLE HEURISTICS

Nathan Coppedge

HEURISTICS USING CATE-GORICAL DEDUCTION

These assume a method using polar opposites in diagonally-opposite positions using the formula AB:CD AND AD:CB, or related formulas for larger sets in a square, Mod 4 form in which the number of deductions is equal to the square root of the number of categories.

(1) Opposites contradict, so non-opposites can be compared directly (with opposites in diagonally opposite positions), creating objectivity and coherence.

(2) Using suspected causal oppositions like (rationalism → post-rationalism, and intuition → exception), the deductions using two such pairs are valid for the specific case, until the case is used up.

(3) One half of an even-numbered category diagram solves the other, when it uses polar opposites in opposite positions.

(4) When AB:CD is correct, AD:CB is also correct, unless the opposites are not balanced.

Nathan Coppedge

INFORMAL CATEGORICAL

1. All it takes is reasonable exaggeration and a little justified context → Knowledge.
2. Methodological relevance → The groundwork of systems qualified by the quality of the application.
3. Original Aesthetics or Useful Philosophy = Big Ideas.
4. Coppedge's Law of Iteration: Quality of iterations increases in proportion to their coherence. Iterations of coherence increase in proportion to the quality of iterations.

PROGRAMMABLE HEURISTICS

PART I

The following systems and heuristics are freely available under the condition that Nathan Coppedge is prominently cited in any resulting implementations, that is, within printed materials and/ or digital descriptive matter viewed with the product, game, or other creation, including within a reasonable time user-visible aspects of all individual resulting software products, whether commercial, non-commercial, academic, etc. Listing (Nathan Coppedge) with other contributors is permitted, but the appellation 'philosopher' is recommended. No royalty payments are necessary for these heuristics, or the duplication of the key parts of these systems. Implementations can claim to be proprietary on the grounds of unique utilization of the systems, but I can not legally guarantee the proprietary aspect of implementations. Likely the advantage of commercial implementations will be the market advantage and utility of the product, whether for predictive power, educational utility, exponentiality (efficiency), larger social welfare (defense, innovation), or some other purpose.

ABBREVIATIONS

Q = Predicting Questions. **Coherent Questions**

OK = Objective Knowledge. **What is categorical deduction?**

P = General Solution to Problems. **What is a general method for solving all paradoxes?**

PPT = Psychic Prediction Techniques. **Forms of Psychic Prediction**

C = Creativity Formula. **Coherent Creativity**

S = Method for Generating the Souls of Literature. **Souls from the Library of Alexandria**

HEURISTICS:

'Q → Psych'.

'Psi → OK → verify'. (*Psi means empirical data*).

'P → Practical'.

'PPT → Data Collaboration'.

'(Metaphor) C Variations Systematized'. And, 'Categories → C → Structure'.

'Informatics → Elaborate → S'.

…

PROGRAMMABLE HEURISTICS AND BIG IDEAS

ADDITIONAL SHORT & MISC

HEURISTICS

Paradigm Shift: A small thing with a large thing or a large thing with a small thing. *[Krishna says 'take with infinite vision'. Socrates says 'he is wisest who knows that he knows nothing'. Religion progresses, then the rise of secularism and the invention of the Calculus. The universe appears to be governed by God, then humanity develops the atom bomb. It is science take all, then some lucky fool invents a new rule…]*

Information: X, Property of X. Is Property of X good or bad? X is [good / bad] because (it is) [Property of X].

1. Based on existing evidence? 2. Provide media attention. 3. Add to list of collaborators. 4. List experiments with total list. 5. Find statistics.

'Proof is Proof. What is proof? Okay.'

Relative to reality X, it must be considered a real X… OR It has no standard X (since standards are relative to reality).

Theory of X → Xish theory.

That involves X → X is a component of the logic.

|Sufficient time for X| : X is occurring.

Insofar as X is smart, and X is Y, Y is smart from the position of X.

The Y of X is the Y of (X def).

'Cheap Bet→ Optimize for 2nd Place'

Influence → Positive Emo → Optimal acceptability.

Can [vernacular] [display] [miracle]? No miracle. Select an applications expert.

Razor of Destruction: Science should be predictable by some logic in retrospect, or it is not logical. If scienve cannot predict universals, because universals are falsifiable, it is not logical. If it is not logical it is not mathematical. If it IS logical, some things which are not falsifiable may be complex and hard to guess. Basically, if we are logical or illogical, there is a problem for science: rejecting mathematics, or rejecting parsimony, but not both.

Razor of Ambiguity: If its not an argument, it is as much an argument that it is or isn't an argument—active in corners of reasoning.

If there's absolutely absolutely no rule which says X is not true, then X tends to be definitely or semantically true.

X to X: circular reasoning, not causal or non-

PROGRAMMABLE HEURISTICS AND BIG IDEAS

causal yet with a pretend corollary.

In principle you could truly be more meaningful.

Manage Fate but never believe in it.

If you don't like a particular time, there may be something you are forgetting or were not informed of about it.

What means something bad, however good in appearance, is opposed by the good insofar as it's bad (for example, we may grow the mind ny thinking the world is made of honey, because babies are like parasites of woman, and babies don't like honey, and the universe concerns the mind, and women are like the universe).

What represents one thing contradicts it when it is weak, and the weakness shows the usual strength only towards the opposite property (for example, if food is given as a symbol of defeated enemies, for example Angeletti pasta. fruit being weak food represents enemies which one has not yet defeated, in essence, potential to defeat enemies, potential advantages over enemies that have not yet been defeated, in essence potential advantages).

What is bad without a bad reaction → and what is bad about a reaction is if it contradicts a primary thing. Evil is contradiction (contradiction of good things).

Pride is a lie is higher-minded than often thought.

Good idea has a detracting property → In the best case, the good idea establishes contradiction with the negative property.

Hard truth(s) are opposite: Measurement of 'devils of difficulty'.

Looking for Next X → Maybe Meta-X (Example: Next physics → Maybe metaphysics).

When there are no immortals, no one has enough magic (similar to Socrates is mortal).

The world may be as dead as you are, modified by sophistication raised to deception. In other words, truth by appearance is as dead as you are unsophisticated.

If life is a challenge you're not supposed to be challenged.

People are motivated by ideas that happen at the time of their prime. I am motivated by 2013, just as Langan is motivated by 1972. Nathan: Think how to solve all problems with a computer. Langan: Think how to out-think computers.

If it is logical to act ethically, then mathematicians act ethically insofar as they are logical, or there is some way math is not associated with

PROGRAMMABLE HEURISTICS AND BIG IDEAS

logic. Escape Strategy 1: It is not logical to act ethically. Escape Strategy 2: Math is not ethical logic.

People who are punished have not been rewarded for analysis.

It could be a problem to edit out extreme examples of things that are survivable and enjoyable.

Semantic technology → Education.

Expensive judgment --> Suspect.

Godgame: Compound of worthwhile things.

It isn't atoms if there's pixellation, if we assume there is not technology creating pixellation. But is it atoms if there's bad vision? Maybe, but I don't know that argument.

The best ideas are like meaning and souls.

How to create a generality: neutral laws.

Solver for Psychological Egoism:

- Arbitrariness may be inconvenience.
- There are preferances with composed volition, there are X values with well-composed X.
- Identity has an internal meaning which must work even without a necessary external meaning, unless external meaning has

an absolute non-arbitrary reason, or if external reality is primary.
- In terms of experience, legends remain legendary, but there may be controls for this.

—<u>What are the implications of psychological egoism, if it's true?</u>

Understanding follows after a scheme or intuition assuming consciousness.

If the context is not absolute, I suppose this value would give 2 - 4 / x proportional units. In other words, as far as the physical world, thought includes merely opposition, synergies, and knowledge.

Those who don't deal with fundamental variables end up drowning in semantics.

Immortal world: We csn hold it in place by making things that don't change.

Physical brain damage + conceptual brain damaged interpretation = stupid.

Medical Heuristics:

Aseptic procedures → Placebo.

And if they are not grand people, then it is not a grand place, which is an interesting principle.

PROGRAMMABLE HEURISTICS AND BIG IDEAS

BACKGROUND SYSTEMS FOR PROGRAMMABLE HEURISTICS PART 1

CATEGORICAL DEDUCTION

[Coherent Systems A.1.A.1.]

Categorical deduction is a method I invented for formulating coherent knowledge using an n-dimensional typology. For most purposes it only operates in four square categories ('quadra') lying inside a bounded Cartesian Coordinate System ('axes').

Deductions are produced when opposite terms or labels, each being of any length, and occupying separate boxes, are arranged to form statements that are said to express all the data that could be expressed----because the words are analogous to everything contained by the concepts.

The words must be opposites opposed along the diagonal, and arranged with an order preference given to categories A and B, which are taken to be the subject of the individual analysis.

Nathan Coppedge

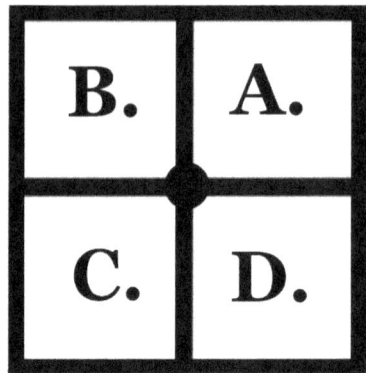

Coherent statements are then expressed as "AB:CD and AD:CB" in terms of A.

Since B and D are switched as part of the operation of the deduction, the preference of B over D is actually unnecessary, although the content is not arbitrary, as it expresses a certain relation of judgment axis B-D with judgment axis A-C.

Examples that don't work:

"Bad women make good men." You may think this is sound reasoning, but it does not logically follow, because if man is the opposite of woman, they cannot both be human. The opposite of human is at least not human.

Examples that do work:

However, you could argue bad is the opposite of good, and the opposite of love is hate, now you can conclude that "bad hate makes good love." That would be logically sound, because it serves as a definition, and we know that it is not contradictory. Of course, other types of exceptions still exist which restrain the ultimate significance both of good and bad, and of love and hate. The statement does not say that it is true in every possible way, but only that it is a logically true definition which measures the extent of validity for that exact case, insofar as the words are accurate representations. If we want it to be measurably true, and not just logically true, then we have to assume that the opposite properties are measurable. And, unless they are absolute, there is no way to be certain that the statement is coherent. However, using similar rules, we can make more complex statements that are equally valid, such as: "good problems with hate produce bad solutions in love." Although more entities are involved, the more complex statements do not have to assume the entities are real or measurable to be logically valid. The deductions also don't depend on the idea of cause and effect, hence the concept of 'non-causal inference'.

Consider the example of beauty versus ugliness, and a sensitive person versus a stoic. We don't argue that any of these entities or qualities don't exist for some person or other. Now, we can't compare opposites directly because that would create a contradiction. So, we compare non-opposites. There is no rule which says that stoics can't be ugly, or that sensitive people can't be beautiful or ugly, etc. In fact, the only thing that would contradict sensitivity is being stoical, and the only thing that would contradict beauty is ugliness. (The only exception to this is irrationality).

Now, we are not saying that stoic and sensitivity or beauty and ugliness cannot be compared to other things, so there is no contradiction in selecting something specific. At this point there is no contradiction. We are comparing non-opposites, because that is not contradictory. It is a possibility, so it can express something about the world. Since the concepts can be defined as the only words that represent the exact same concept, or the identical concepts are interchangeable and have only one opposite, therefore the comparison of non-opposites represents the only available knowledge on whatever topic the terms concern. Since it is the only available knowledge, it is the best knowledge, and where opposite terms are exclusive of all possible descriptions, it is also universal knowledge. Now it follows that 'a beautiful stoic is ugly sensitively' and, under different conditions, 'an ugly stoic is beauty-sensitive'. Otherwise, the terms are not opposites, the process is resolved differently, the interpretation is naive realist, or there is a paradox, irrationality, or incoherence. In sheerly objective terms the only exceptions I know of are naive realism, paradoxes, irrationality, and incoherence. There are also several other trivial exceptions, such as when formal tools are rejected out of hand, or when an informal tool is assumed to be more important.

What is potentially unique about the system is not just its sense of double relativism which I call relative absoluteness, but the way it works across language, and for any extreme concept.

For assumptions of the system, SEE: Nathan

PROGRAMMABLE HEURISTICS AND BIG IDEAS

<u>Coppedge's answer to Can philosophy be axiomatized?</u>

However, things like 'cat and dog' or 'man and woman' don't work except in what is called a 'modal sense'. The modal sense is the same sense as 'this lamp post to that street over there' --- it may not in fact be opposite. However, terms like cat and dog can be lumped into categories like animal and human, and opposites can be imagined for them, like 'dead human' and maybe 'nativity water' for 'alien flame' or the like. These sorts of concepts at least set up a relationship for logical comparison coherently, providing a meaningful standard for correspondence that was not previously imaginable outside of science fiction and Alien Phenomenology.

TOE Over-Unity Rating: 1

—<u>Over-Unity Formula for TOEs</u>

GENERAL SOLUTION TO PARADOXES:

[Coherent Systems A.1.A.2.]

The method called paroxysm or double-paradox is to my knowledge the first and only proposition ever made for the solution of all paradoxes.

It is a method I formulated first in 2014, as a result of categorical developments in a book I was writing called **The Dimensional Philosopher's Toolkit**, a self-published work.

The method of paroxysm involves choosing the opposite of EVERY WORD in the best definition of the original paradox (except the word 'paradox' which is not included), and combining them in the same order as the respective original words. The result serves as a solution to the paradox.

So, for example, with the Sorites paradox of the sound of hay falling, if we define the problem as "definite continuum" the paroxysm becomes "indefinite definitions". If we define the problem as "meaningless continuum" the paroxysm is "meaningful divisions". Depending on the way the problem is defined, it has different solutions. This is the only way to be sure that the paradox is solvable.

For more information on Paroxysm,

PROGRAMMABLE HEURISTICS AND BIG IDEAS

see: <u>Paroxysm (Formal and Logical systems)</u>

Note: *the method also extends to solving a wide variety of problems, in the sense that it defines what remains to be done. But this involves placing the term 'problem' with the problem, and 'solution' with the solution, because paradoxes are the only formal problems that remain problems when they are solved, and thus under my definition, are the only types of universal problems except in an incoherent universe. An example of a problem as opposed to a paradox is: problematic war is solved by peace. Problematic hate is solved by love. Notice the similarity to my method of categorical deduction. Categorical deduction is a more general method, since it deals with a wide variety of word pairs, not just 'problem vs. solution' and the opposite of a problematic term.*

See also a high-minded variation: <u>High-Minded Colundrum by Nathan Coppedge on Official Nathan Coppedge Blog</u>

TOE Over-Unity Rating: 1

—<u>Over-Unity Formula for TOEs</u>

FORMS OF PSYCHIC PREDICTION:

[Coherent Systems A.1.A.3.]

PSYCHIC HINTS:

1. There is something similar about it to something important to you, for example if you use a Q-tip for your skin medication, you might see a Q-tip on the ground within 3 days to a week, *that wasn't dropped by you.*
2. Something seems especially familiar like you keep thinking about it day after day, then it changes or something happens to it.
3. Something has been creeping in the back of your mind, like it is creepily familiar. For example, you can't stop thinking of car accidents but it never completely surfaces in your mind, like someone is trying to hide the information. Then you witness a car accident.
4. Something draws your attention the first time you see it, then someone uses it or it has an intimate role to play (in love it may be the opposite: things that don't seem to matter come into play).
5. Something resists your attention, then you think of something strangely related to it, then the strsngely related thing starts to happen immediately, much later, strongly, etc.

PROGRAMMABLE HEURISTICS AND BIG IDEAS

A prereq for parts of this are: Poetic Planning : "What is appropriate will happen, accompanied by the response to reality which makes it so."

Followed by a statement of what has come before this reality, which, when stated as a prophecy predicts what comes after.

PSYCHIC CALCULUS

> Michael's Formula for the Psychic Calculus: The Opposite Limitedness, in Opposite Order if Necessary.
>
> Take an example:
>
> Sweden, say it means a place where they like to give the Nobel Prize, the psychic prediction is that no one wins the Nobel Prize.
>
> A popsicle is frozen as long as it doesn't melt, and gives a sweet feeling, the psychic prediction is that there is a big ball of fire, and the popsicle melts, and you get a bitter feeling.
>
> A chai tea from Starbux is the perfect drink, the psychic prediction is it's better if you don't drink it, or it's perfectly ugly for you because others don't drink it.
>
> You're wearing sweat pants because you expect to sweat. The answer is you'll need them when you don't expect it's cold.
>
> —The Psychic Calculus

PSYCHIC PREDICTION TECHNIQUES

Psychic prediction may take several basic forms.

First I will describe the most basic types of prediction.

First of all, the most basic type is 0-dimensional prediction. This consists of predicting what has already occurred, that is, predicting the types of things that have already happened. A second degree of this is had by predicting things that are similar to those things that have happened. For our purposes, this can be called simple generalization. *If Henrick usually wants to play games, perhaps he wants to play games now*. This is the first dimension of prediction, and it is the type that gains most easily by probabilistic inductions. This method is also called specialized prediction when it is applied to specialized modes of behavior. For example, we can predict that a Matisse will sell high compared to an unknown artist. We know that popular items in an auction sell high, whereas unpopular items might not sell at all. Therefore, there is an exponential relationship for example, between selling a Matisse, and selling a Matisse at an auction. These kinds of things can be predicted by studying the specific character of the modalities and events involved in a given situation. However, if an event is instead informal or contrived, this lends an aspect of unpredictabil-

PROGRAMMABLE HEURISTICS AND BIG IDEAS

ity. The predictions only work when all of the prior conditions are met, and become less predictable with every difference from the previous cases. Therefore, differences can be used to predict differences, as another type of specialized prediction. It may help to predict trickery or confusion ('likely outcomes'), rather than predicting a specific event. It should be accepted that some conditions and choices are arbitrary. *Because we do not know if conditions will be met to satisfaction, we know that some events are arbitrary*. If the conditions are one half different, then prediction requires a strong degree of formalism, however that is calculated. It involves, in effect, exceeding expectations, or coming across an event that happened *just in the same way*, but as if by chance. This is one reason that scientists have been known to require the reproduction of *laboratory conditions*, even with highly predictable phenomena. Thus, specialized predictions have some limitations.

The next type is delineative or elaborative prediction. What it consists of is a generalization modified by additional imagination about the significance of the factors involved. This type of prediction can be called variablistic, because it often functions by applying a generalization to a deduction about a variable. *If elephants are painted red, perhaps it is a sight for sore eyes*, etc. One form of this is prediction through emergence. This is not necessarily a linear prediction because it essentially doesn't predict based on existing data. Nor does it predict based on known exterior data. Instead, it involves a conclusion that *something is missing from the data*. Logical conclusions are drawn so that we can make major systemic conclusions about what the data means. The new theory appears *as if from thin air*. This is similar to the emergence of Darwinism, or the genetic explanation of reproduction. What determines the success of these theories is their relative importance, not necessarily the lack of any alternative. It is the importance of the theory---- its *emergence*----which drives the prediction. (Many theories from social science involve emergent theories, such as socialism and capitalism. Instead of acting as a formal constraint, they often expand the way that the conditions function. In this case, the explanation is not erroneous, but instead, serves as a *new rational mode of explanation*).

A third type is contingent or categorical prediction. If something is the case, then we can predict that the things that rely upon this first condition are modified when that category is modified. This form of prediction works better for predicting quality differences than actually-different conditions. However, if multiple qualities are absent, predictions can be made about the alternatives. *If there is no snow, it can be predicted that it is not cold, or there is a shortage of water, for instance. If it is not cold, one can predict that it is arid or moist. If there is a shortage of water, one can predict that it is dry, or there is a high tolerance for water.* This can also take the form of complex categorization. Attaching variables to a given object means that predicting the outcome for

PROGRAMMABLE HEURISTICS AND BIG IDEAS

the main object affects the outcome of some, if not all, of the variables. For example, '*if we do something extreme, the change might be observable. Otherwise, it is an abstract or unmeasurable form of extremity. We must have some means of observation, or we can usually conclude that the effects are not extreme. Or we can adopt an irrational view*'.

A fourth type of prediction is coherent prediction. This is also called synergism or epiphany. The simplest form of coherent prediction occurs by the exclusion of all but one unlikely option. *Hella spent a hot day in the desert, and she was outdoors, and walked several miles, time passed and she didn't expire: she must have brought something to drink with her.* A more complex form occurs by qualifying what it means to make a given combination. *People who have complicit sex are always lovers. Therefore, if two people have sex, it might be complicit, and they might be lovers. Or, something is complicit between two people. If it is sex, they are lovers.* This can even involve highly complex phenomena. For example: *Joe defines himself as an editor, but he works as an economist. In some way he is doing economic editing.* This is the beginning of a genuinely psychic method. Attaching judgments of fully embraced variables can be a meaningful way of reaching for epiphanies. For example, what 'definitely IS something' about a given thing? Then apply that condition to factors like responsibility, organization, and predictability. An exception to this is so-called 'black swans'. In that case, one must predict the rationale which *makes* something a black swan. The rule in that case is that things are either unreasonable, reasonable, without purpose, or serving a prescribed function. A method for solving black swans involves corroboration or defaulting. This occurs when there is no better explanation remaining for a given thing. *Well, we know that such-and-such a creature has eyes based on the related species, but nothing about the creature looks exactly like eyes. The eyes must be these spots on its back. Otherwise its blind.* Or, black swans could exist, as long as we know that color serves no inherent function.

Now for more genuine psychic predictions:

A second genuine form of psychic prediction involves using a posteriori reasoning on a 0-dimensional prediction. For example, *if we know that some events are arbitrary, then we can derive that we don't know if some conditions will be met to satisfaction. If we know Henrick wants to play games now, we can predict that he usually wants to play games.* This form of prediction often involves deducing the types of statements that lead to a particular line of reasoning: that is, *predicting a rationale.* Many psychics are familiar with this way of phrasing deductions.

A third form of genuine psychic prediction involves determination based on unstated facts. Since everyone thinks about the opposite of what they say, at least unconsciously, combining multiple opposite terms for terms that have been stated as someone's opinion, or as the definition of a motive or interest for the person or organization, will give information about the genuine motivations, or else the looming unknowns in the life of the person or organization. For example, if someone states that the first thing on their mind is their motorcycle, and the second thing on their mind is their manhood, then you can predict that they're concerned about meeting someone else on a motorcycle.

A fourth form of genuine psychic prediction involves categorical relationships. One can ask or predict 'what is someone's usual mode of relation with the world?' Then one can predict that they use that mode of relation with their perceived opposites. For example, an artist who expresses that the thing on his mind is cars can be predicted '*not to buy a painting of a car, instead you'll make it yourself*' (the concealed opposition is between the artist who makes art, and his opposite, the buyer of the art. The opposite of making a painting of a car is buying a painting of a car). Similarly, if a business expresses itself as aggressive and competitive, but you think they're liars, you can predict they'll have contradictory marketing ('competing truths', since their mode of relation is competition, and their opposite is the truth).

A fifth form of genuine psychic prediction: take any number of factors describing a current event or situation you're in, and reverse the factors that are different from the subject. This can be used to predict how someone is feeling, or what their core motivation are. For example, an artist is at a business convention. So they're feeling unconventional, and they feel like making art, since that is not a different motive from business. Or, a philosophy society is at an art gallery. So, it thinks its popular art ('society' does not conflict with 'gallery'), and it thinks its un-philosophical art, or tries to make connections between art and philosophy ('philosophy' is different from 'art' or it can be debated). Other conclusions might be that they think art is trying to commercialize philosophy, that philosophy ought to involve graphics, or to view art or philosophy as a socialist movement.

The sixth form: Coherent Questions. The best formula I have found for what someone desires to know is: "The opposite of you related to something different from both your opposite and you." For example, someone sadistic may want to know "What happens" (to a victim). A philosopher may want to know the reason for practical things. A child may want to know who they would be when they grow up. (An old person may want to know when they will be young again, or when it will be over).

The seventh form: good X → good X (Y). Good Y → good Y (X). Moderate X (Y) and Y (X) = medium of prediction.

Those are the ten categories of prediction that I have determined. I hope this writing may be considered useful to my readers on this most

PROGRAMMABLE HEURISTICS AND BIG IDEAS

often unrealized subject.

Coppedge, Nathan. *The Dimensional Psychologist's Toolkit*. CSIP, 2014

CRITIQUES:

1. Someone offers a psychic model.
2. Someone offers a psychic model about them.
    If BOTH fail to make *accurate* predictions, *then what*? No more psychic model, possibly!

—<u>Accompaniment for Correlative Reasoning and Causal Inference</u>

Nathan Coppedge

COHERENT CREATIVITY:

[Coherent Systems D.2.D.2.]

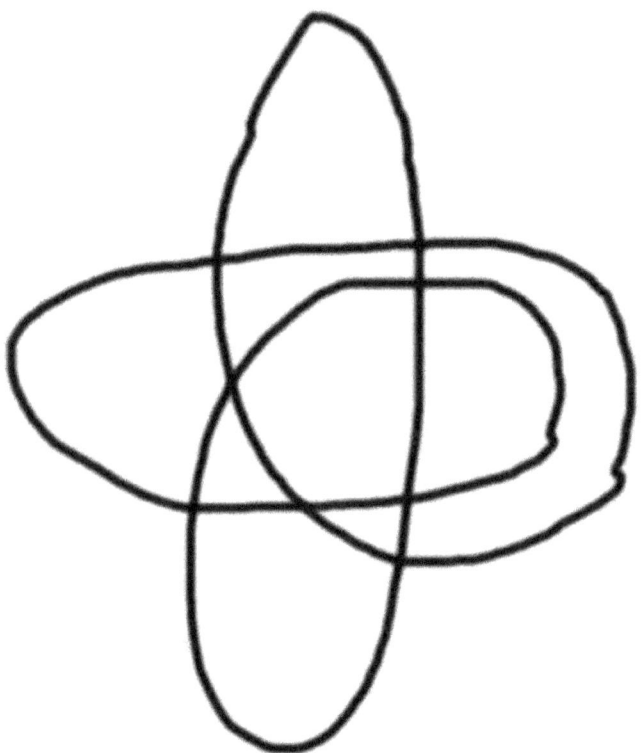

An attempt to describe the coherent formula for creativity…

PROGRAMMABLE HEURISTICS AND BIG IDEAS

Next Step: Angle of escape from one neighboring list of categories into another. E.g. coherent angle is the creative angle distributed into the ordered sections of an exclusive list. Modular angle is the same, but with a non-exclusive list. Explorative angle is the same but within a larger creative environment like canvas or fully or partially-mapped proportions of importance. Quadratic escape is same angle into quadratic list of categories. To eclipse next list is to create an organizing angle equal to the angle minus the number of degrees expressed by the quadratic categories in the next list, or to instantaneously fully process and re-apply the category the angle falls into before continuing process.

Ultimately: Contingent 3-d.

[Originally posted March 23, 2017].

Here is a paradox: how do we explain the Fibonacci sequence exclusively in terms of pi (the circumference of a circle with the diameter 1)? The answer is that we can't, because as soon as Fibonacci is all about Pi, Pi no longer describes all of Fibonacci! So, how could a mathematical theory ever achieve a theory of everything? My answer is that there is more than one theory of everything, and many of them will not be mathematical. —An answer on a topic by Coppedge

FORMULA FOR SOULS:

Soul of the book =

'If you [X] qualifier [subject of X and qualifier] [opp X clarified]'

Optionally, you can add a moral:

'The [subject of X alone] is [verb / adj. of opp qualifier]'.

Note: For this second part, it may help to refer to the below title
formula, in order to find the opp qualifier before it is modified.

Title of book =

Usually: '[quality of X] [opp qualifier]'
For perusal, or quick use: '[opp qualifier] [quality of X]'

TOE Over-Unity Rating: 11

—<u>Over-Unity Formula for TOEs</u>

PROGRAMMABLE HEURISTICS AND BIG IDEAS

The easiest way to use the formula is to generate original souls
and then find the corresponding titles by finding the most
essential, knowledgeable quality of X and then finding the
opposite of the qualifier introduced in the soul of the book.

Together with the soul the title provides a basic index of the
value of any text, and permits exponentially efficient reading.

For example,

Soul = 'If you die early enough you live'.
Optional 1= 'Death is the aging process'.

Title of book = 'Bad Archaic'
Bad [= qual. of die] Archaic [=opp of early]

Another example chosen more arbitrarily,

Soul = 'If you live surely truth may die'
Optional 1= 'The life is uncertain'.

Title of book = 'Optimal Uncertainty'
Optimal [=qual. of live] Uncertainty [= opp

of surely]
Also, optionally [added after optional1],

Optional 2= '[A/the state / process from verb / adj.] of / is
[property of qual. of X]'

Optional 3=
A.
[2nd part of optional 2] [1st part of optional 2] [opp qual.]

B.
High form of quality (only included in Poor Zurid so far)

End=
[(Desire such as at last / as it happens / or quality of X association)... (Unobvious).
(Unobvious)
State = Knowledge /
Process = is /
Pure and partial processes = and this will continue /
State process and state-as-process = that's what we think /
Capital states and qualities = title /
Object = it's like (knowledge of object) referring to 6 possible conditions of the first part of Optional2]

PROGRAMMABLE HEURISTICS AND BIG IDEAS

FORMULA FOR PERFECT QUESTIONS
Updated May 9, 2019. Formula from 2018 by Sept 11.

[Coherent Systems D.2.C.1.]

THE MOST IMPORTANT QUESTION

- *There is a means to guess the ideal question pertaining to any given thing if we know what it is concerned with, and if we apply that concern to a more general concern involving the original subject.*

It is the question of:

What question is unimportant for [my polar opposite]?

For example, if you are philosophical, it is a question like, "Why wouldn't that guy do philosophy?"

If you are very early in your life, it is a question like "Why can't adults be like kids?"

If you are very old, but led an unwise life, it is something like: "Why are young people so good at making decisions?"

If you are a powerful person who just got married, it is something like "Why are single people such weak personalities?" or "What prevents them from taking her away?"

If someone is old and wise, the proper question is something like: "Why are the young people unconcerned with wisdom?"

GENERAL FORMULA

The best formula I have found for what someone desires to know is:

"The opposite of you related to something different from both your opposite and you."

For example, someone sadistic may want to know "What happens" (to a victim). A philosopher may want to know the reason for practical things. A child may want to know who they would be when they grow up.

(An old person may want to know when they will be young again, or when it will be over).

TOE Over-Unity Rating: 0

—<u>Over-Unity Formula for TOEs</u>

PROGRAMMABLE HEURISTICS AND BIG IDEAS

FORMULA FOR ANSWERING ALL QUESTIONS

FORMULA FOR ANSWERING ALL QUESTIONS:

1   *Significant updates on Sept 11, 2018, approximate date of Answer Formula.*

2   [Coherent Systems 2.B.1.D.1.]

3   To give a satisfactory answer to any question:

4. First, properly formulate the question. <u>Coherent Questions</u>
1. The answer is formulated by saying the opposite subject of what you are concerned with in the perfect question confirms the object of the question.

Some credit to my Dad Michael Coppedge for this answer.

If you are philosophical, the question is , "Why wouldn't that guy do philosophy?" and the answer is, "I'm a philosopher." or, "That guy is more philosophical than me, I can be more philosophical than him."

If you are very early in your life, it is a question like "Why can't adults be like kids?" and the answer is "I've had an epiphany, kids can be like adults!"

If you are very old, but led an unwise life, it is something like: "Why are young people so good at making decisions?" And the answer is "If I'm bad at making decisions why don't I just be old."

If you are a powerful person who just got married, it is something like "Why are single people such weak personalities?" or "What prevents them from taking her away?" And the answer is "Ugly people have strong personalities" , "Strong people are impersonal" or "She has a lot of personality, but her ugliness blows me away" or "It would be ugly if you blew me away."

If someone is old and wise, the proper question is something like: "Why are the young people unconcerned with wisdom?" And the answer is: "The concern of wisdom is getting old." or, "Foolishness is not young in this world." alternately the question is : "Why are young people so blind?" and the answer is: "I must recapture my sight!"

TOE Over-Unity Rating: 3

—<u>Over-Unity Formula for TOEs</u>

PROGRAMMABLE HEURISTICS AND BIG IDEAS

WIZARD LOGIC:

Sept 14, 2018

QUICK VERSION: Negative possibility qualified, Otherwise, opposite slightly unqualified.

…

Things will probably by crumby if I am not divine in some way.

There will be a warning if we have no epiphany.

Silence will contain it if we do not magnify it's power.

The properties are myriad that do not contain death.

Are we certain of this—it's entirely optional?

Have we broken the lock that contains the key?

The pattern that is impassable allows us to see!

The enormous danger of moving images is they do not capture life's small wonders!

The ticket—as far as we can see—is they have no limitations, which they do not give.

(Qualification) of (negative possibility) if not (positive possibility) (slightly opposite qualification).

…

FURTHER EXAMPLES:

Is magic death, or is death merely the mirror-image of life?

I have tolerance for sugar rain, because it wipes away the sugar.

…

TOE Over-Unity Rating: 4 ($5^1 > 2 - 1$)

---Over-Unity Formula for TOEs

POWERFUL ALTERNATE FORMULA:

Real anti-magic is still magic if it's impossible. For example, an impossible inquisitor.

LAW OF THE GODS:

When there is no obvious explanation, it is some type of influence that is in play. Influ-

PROGRAMMABLE HEURISTICS AND BIG IDEAS

ences are gods, as any true influence is magic.

ALTERNATE WIZARD:

(Obvious statement, then) a mysterious ironic thing, followed by an unexpected negative reference (to magic), followed by a certain contradiction.

Examples:

Cursing is the worst. I think it's the opposite of cursing to be cursed, the question is how to be cursed in the right way?

One could try negation. It seems quite strange... how was that true? Falsehood has never been so obvious!

It's spelled differently... Whatever it was, it still is... it is not the same as certain thoughts... what is the same is still different.

Why problems? A problem-nature has that form... spirited at will... it's spirit has a contradictory solution!

A mysterious notion... The most enigmatic ideas are that thing... taken away by magic... returning into obscurity.

An act of Fate... The norns are quite willful, so we are uncertain of the outcome.

A trist... The pattern has been broken, so at the

end things will come together.

Stupifying idea… What we have said is not all as we thought, so we are certain of a foolish outcome.

A dubious outcome… Are you sure of yourself? One might expect discordances, only they never last!

Be certain you have freedom…. What tool? Tools are for the maimed, therefore we don't use our tools.

Wide array, generalized: <u>Objective: Wizard Logic</u>

Generally…

$|2 / D > results / verbs| = $ Possibility equation.

$1 / $ (ratio of numbers from equation) $=$ new number.

PROGRAMMABLE HEURISTICS AND BIG IDEAS

Now with magic...

Standard category formula for 10 =

1 / Math (0.1) per 1 = 10

(Version 1)

1 result, 2 verbs, 5 dimensions

2 / 5 > 1/2 = 4/10 > 5/10 = |-1/10|

1 / (1/10) = 10 (values were for the original Wizard Logic)

...

How to get other wizard logics?

Keep results half of verbs...

Dimensions has to be 5...

Only other option is changing results and verbs in relation to different dimension...

Restraint is placed because verbs cannot be less than results.

Another option is a 4/20 relation between

results and verbs in 5D or a 2/5 relation between results and verbs in 4D, and other relations exist in higher dimensions.

(ANOTHER VERSION)

1 result, 10 verbs, 2 dimensions =

2 / 2 D > 1 / 10 = 1/10

1 / |1/10| = 10 (Wizard Logic)

ANOTHER VERSION

2 results 4 verbs, 5 D

2/5 > 2/4 ... 8/20 > 10/20...

1 / |-2/20| = 1 / |-1/10|

= 10 (Wizard Logic)

ANOTHER VERSION

3 results 6 verbs, 5 D

2/5 > 3/6... 12 / 30 > 15 / 30...

PROGRAMMABLE HEURISTICS AND BIG IDEAS

1 / |-3 / 30| = 1 / |-1/10|

= 10 (Wizard Logic)

ANOTHER VERSION

9 results 10 verbs, 2 D

2 / 2 = 1

1 > (9 /10) = 1/10

1 / |1/10|

= 10 (Wizard Logic)

ANOTHER VERSION

3 results 10 verbs, 5 D

2/5 > (3 /10) = 4/10 > 3/10 = |1/10|

1 / |1/10|

= 10 (Wizard Logic)

ANOTHER VERSION

6 results 20 verbs, 5 D

2/5 > (6 /20) = 8/20 > 6/20 = |2/20|

1 / |2/20|

= 10 (Wizard Logic)

ANOTHER VERSION

2 results 5 verbs, 4 D

2/4 > (2 /5) = 10/20 > 8/20 = |-2/20|

1 / |-2/20|

= 10 (Wizard Logic)

PROGRAMMABLE HEURISTICS AND BIG IDEAS

WIZARD LOGIC CHART

D, Results, Verbs, Rating

2 D = 9 results 10 verbs = 10 (Wizard Logic)

2 D = 1 result 10 verbs = 10 (Wizard Logic)

4D = 2 results 5 verbs = 10 (Wizard Logic)

4D = 4 results 10 verbs = 10 (Wizard Logic)

5 D = 1 result 2 verbs = 10 (Wizard Logic)

5 D = 2 results 4 verbs = 10 (Wizard Logic)

5 D = 3 results 6 verbs = 10 (Wizard Logic)

5D = 3 results 10 verbs = 10 (Wizard Logic)

5 D = 4 results 8 verbs = 10 (Wizard Logic)

5 D = 5 results 10 verbs = 10 (Wizard Logic)

5 D = 6 results 20 verbs = 10 (Wizard Logic)

Notes: When a value = 2 it is suggested classic wizardry or categorical deduction are involved somehow, when a value = 4 or 5 it is suggested it means 4 or 5 dimensions or categories are separately involved somehow.

PROGRAMMABLE HEURISTICS AND BIG IDEAS

INCOHERENT DEDUCTION:

LOGIC

In an incoherent place, if

X looks opp X, then opp X doesn't look X.

Otherwise, coherence applies.

For Ex,

If we observe art and the good looks bad, then the bad also won't look good. Otherwise, there is grounds for a universal rule about good, bad, or both.

> TOE Over-Unity Rating: 0
>
> —<u>Over-Unity Formula for TOEs</u>

…

NOTES ON INCOHERENT LOGIC:

In a coherent place, if the good looks good, and the bad looks good, that is universally good. Likewise, if the good doesn't look good, but the bad looks bad, then everything is bad. However, if the good looks objectively bad and the bad looks objectively good, this means objectively that things look like their opposites. So the only case where incoherence applies is when one comparison looks opposite and one does not as far as a particular judgment. An exception to this which might not be incoherent is when everything is made of time, and categories are inter-deferred, like Kripke's theory. It may be that Kripke thought this was the wave of the future, as it connotes technology, although in my theory coherence might be highly useful, and we already know incoherence is useful for science. Thus what emerges, as far as incoherence, is a three-pronged approach similar to Hume's Fork: one part analytic, one part scientific, and one part radical technology.

…

HOW TO DETERMINE INCOHERENCE IN A COMPUTER PROGRAM:

(1) If there is only one category, (2) If the total categories are not composed collectively of only true polar opposite pairs, (3) If the categories do not fit a coherent formula, (4) If coherent formulas do not apply.

PROGRAMMABLE HEURISTICS AND BIG IDEAS

THE THEORY OF EVERYTHING:

June 26, 2019

<u>Collected Inventions of Nathan Coppedge</u>
[Coherent Systems 2.B.2.C.3.]

INITIAL CRITIQUE OF THEORIES OF EVERYTHING

Nicholas Rescher writes of properties which contradict a theory

There is a:

Principle of sufficient reason

$\forall t \exists t'(t' \, E \, t) \forall t \exists t'(t' \, E \, t)$

where E predicates explanation, so that $t' \, E \, t$ denotes "t' explains t".

And a:

Comprehensiveness

$\forall t(T* \, E \, t) \forall t(T* \, E \, t)$

And a:

Finality

Finality says that as an "ultimate theory", $T*$ has no deeper explanation:

$$\forall t((t\ E\ T*)\rightarrow(t=T*))\forall t((t\ E\ T*)\rightarrow(t=T*))$$

so that the only conceivable explanation of *T** is *T** itself.

And a:

Noncircularity

$$\nexists(t\ E\ t)\nexists t(t\ E\ t)$$

Finally, usually there is a:

Impasse

The impasse is… that… comprehensiveness and finality… conflict with the fundamental principle of noncircularity. A comprehensive theory which explains everything must explain itself, and a final theory which has no deeper explanation must, by the principle of sufficient reason, have *some* explanation; consequently it too must be self-explanatory… how, he asks, can a theory adequately substantiate itself?

Theory of everything (philosophy) - Wikipedia

RECONSIDERATION:

Some years ago I began to get the feeling that equations which are high-minded tend to use efficiency as one of the terms.

Little did I know this would be the basis for a theory of everything.

PROGRAMMABLE HEURISTICS AND BIG IDEAS

Such a theory would have different properties than the above:

(1) It would be sufficiently general, therefore it would have externality, CONTRARY TO CIRCULARITY.

(2) It would have acceptable maximum efficiency, and would be open to critique, therefore it would be contingently irreplaceable, so it would be AMENABLE, QUA FINAL.

(3) It's scope would be so broad that it has a diminishing probability of not describing all things past a certain degree of importance, therefore it has an EXCELSIOR IRREGULARITY OR COMPREHENSIVENESS.

(4) It must achieve it's success with extreme consistency and perfection, so to speak flawlessly, which is to say it has an ability to predict successful paradigms, MAKE PARADIGMATIC PREDICTIONS, SO IT HAS SUFFICIENT REASON.

The typical problem is not that these properties are circular, but that they are extremely difficult.

Now there is an answer.

For my answer I focused especially on systems behaviors.

THEOREM OF EVERYTHING:

Set 0 > Efficiency* + Difference

*Where Efficiency is greater than 1 if object is acting and less than 1 if object is being acted on.

EXAMPLES:

Perpetual Motion:

Required mass > 50% + difference in mass.

Objective Knowledge:

Coherence selects two comparisons that are not 100% opposite.

World Peace:

What we should do, is do less, unless we know what to do.

Souls:

A soul has four parts if the name has two and the difference is two. A typical name has two parts. The difference is two because a full description of two options involves four categories (the name is being treated as the efficiency). To have a value of less than four, the efficiency and difference should result in sections of 1, 0.5, 0.25, and 0, thus the soul is defined by these numbers. Since the difference is always two, each of these values will be multiplied by 2 to equal their efficiency. The first value is 2, which means identical to the first part of the name. The second part is 1, which means the contrary of the second part of the name, the third part is 0.5 which means the result of 2 upon 1. And the fourth part is twice nothing, means reflecting on the opposite of the first part. (This is a direct match with my soul formula found through an independent method).

Women:

What she prefers matters because she's her.

Men:

Super-efficient, more than what I do, but I don't do enough.

Neutral:

No friction.

—Anything Theorem

...

KEY EXAMPLE:

If the Topic is Active, you want to maximize the Concern.

If the Topic is Passive you want to be different than the dominant concern.

Peach Ice cream that is about peaches should add more peaches AND be similar to the competition.

Peach flavor that is mostly about ice cream should add more flavors.

For physics,

Matter that is about energy, with matter being primary, should maximize matter to maximize energy = gravity theory.

Matter that is about energy with energy being acted on by matter should instead concern high-energy states, in which case physics is maximized when matter and energy are almost the same, which is like information theory.

PROGRAMMABLE HEURISTICS AND BIG IDEAS

> On the other hand, if the imperative is to find a theory greater than physics, you should always look outside physics.
>
> But, if physics is really great, you want to remember the greatest physicist.
>
> —<u>Valuable Notes on the Theory of Everything</u>

Some credit to John Miller, Brian Coppedge, Yan Yang, Michael Coppedge, Edmund Scarpa, and Jonathan Berkowitz (in addition to Nathan Coppedge) *for contributing to the theory.*

Soul: Analogous Functions.

    TOE Over-Unity Rating: 9

—<u>Over-Unity Formula for TOEs</u>

[END OF PRIMARY PROGRAMMABLE HEURISTICS BACKGROUNDS]

Nathan Coppedge

PROGRAMMABLE HEURISTICS AND BIG IDEAS

GODLIKE SYSTEMS

SENSITIVE WEIGHING OF SOULS

Jan 18, 2018.

If I can tell my unborn son that he is great, while he still lives in my body in some form, there is nothing countermanding this—

As long as he is open to learning what there is to absorb.

In fact, contradicting the honesty of learning contradicts the senses—it is just a matter of making sure that I am sensitive enough, and that I mean what I say, and I may know the subtlty of perception that accompanies the first thoughts.

It is possible.

There are manifold things.

You matter.

MORALS:

Neither real nor anti-real.

Loose and open.

Brainy 'like a prostitute'.

Measurements of life and death.

The holy truth that everything matters.

TOE SENSITIVITY FORMULA

2.16333 / (<u>invent number</u> minus 100 inverse)

...

The Artist 100 = Infinity.

110 = 2.16333.

101 = Sometimes 2.16333.

120 = 0.9245

102 = Sometimes 0.9245

130 = 0.721111

103 = Sometimes 0.721111

140 = 0.5408333

104 = Sometimes 0.5408333

150 = 0.4326666

105 = Sometimes 0.4326666

PROGRAMMABLE HEURISTICS AND BIG IDEAS

160 = 0.3605555

106 = Sometimes 0.3605555

170 = 0.3090476

107 = Sometimes 0.3090476

180 = 0.270416666

108 = Sometimes 0.270416666

190 = 0.24037037037

109 = Sometimes 0.24037037037

101 lower = 0.2163333

102 lower = 0.10816666

103 lower = 0.07211111

104 lower = 0.054083333

105 lower = 0.04326666

106 lower = 0.03605555

107 lower = 0.03090476

108 lower= 0.0270416666

177 = 0.0280952 (The Experience Machine).

168 = 0.025155 (Perpetual motion ma-

chines).

109 lower = 0.024037037

254 = 0.004796747 (Nirvana)

255 = 0.003926 (winnebago illusion)

[-0.02998: lost energy].

Accounting 2 = 0.024307- (negative)

Impossibility 1 = 0.02185185

PROGRAMMABLE HEURISTICS AND BIG IDEAS

HOW TO ACHIEVE NIRVANA:

June 7, 2019.

254 ——(flip of previous) NIRVANA ENLIGHTENMENT IDEA ETC.:

Impossible? No.

Solved yet? No.

Precedents? No.

Practical? Yes.

Worth it? No.

Extreme? Yes.

Devilish? No.

Difficult? No.

…

A REALM UNTO ITS OWN SACRED PRE-QUEL…

…

…

LIKE CALCULUS, BUT QUIETER.

…

…

THIS SACK OF RAGS WOULD BE PRETTY TO EXPLORE!

WOULDN'T SOMEONE DESIRE THIS IF THEY WERE TOLD!

BUT IT IS NOT EVEN A TRICK, IS IT?

…

WHAT IF IT WERE OF A PETTY KIND, WHO WOULD CARE?

IT IS LIKE SAMADHI WITH ADDITIONAL INSTRUCTIONS!

…

WHO MAKES IT OUT DOES NOT DIE IN VAIN---PETTY INSTRUCTIONS.

…

I THINK I HAVE NOT SOLVED THE CLOUD.

THE CLOUD IS IN ME AND I AM RE-LATED.

PROGRAMMABLE HEURISTICS AND BIG IDEAS

WHAT WATER IT GIVES ME IS PLENTY---

SO GOES THE CLOUD SO GOES ME---

ETERNITY IS LIKE NIRVANA!

IT IS INTELLIGENT HUMILITY

A CERTAIN INSIGHT---

OPENING!

INTO NOTHINGNESS!

A TERRIBLE SECRET FORESEEING NOTHING!

EMPTINESS IS NOT SO SACRED!

PICTURES ARE MORE EMPTY!

IT ATTRACTS GOOD MINDS!

IT KEEPS IT'S TERRIBLE VIRTUE!

SO SAYETH NIRVANA TO ME!

NIRVANA IS A GODDESS WHO IS THE SAME PERSON.

BEING THE SAME, SHE USES THE SAME WORDS.

I AM LESS THAN HER, BUT I SPEAK THE

Nathan Coppedge

SAME WAY SOMETIMES.

OTHER TIMES I FALL LIKE RAIN FROM THE CLOUD.

WHEN I HOLD HER WORDS IN MY MOUTH THE CLOUD…

Notes:

Considering anythings many times.

PROGRAMMABLE HEURISTICS AND BIG IDEAS

HOW TO CREATE MATTER:

June 7, 2019.

1

39 DRAMAS:

Impossible? Yes.

Solved yet? Yes.

Precedents? Yes.

Practical? No.

Worth it? Yes.

Extreme? Yes.

Devilish? Yes.

Difficult? No.

EXAMPLES:

How it must have seemed…

Such pittances and postules.

2

Knowledge of isocreme.

EXAMPLES: Space, colored space.

HYPOTHETICAL MAGIC TORNADO:

Wormhole mouths, like anything else, follow the geodesics of a spacetime.

In the absence of significantly altering the larger wormhole stress-energy, the smaller wormhole mouth would simply move along the geometry and act as a branch point inside the larger wormhole throat, resulting in a multi-throated wormhole.

Altering the stress-energy distribution would, depending on the specifics of course, collapse the larger wormhole into a black hole.

—https://www.quora.com/What-would-happen-if-a-wormhole-were-to-go-through-a-different-larger-wormhole/answers/151262009

PROGRAMMABLE HEURISTICS AND BIG IDEAS

LOGIC OF POETRY:

June 19, 2019

…

Something makes me think it is Y not X.

Let us sing of X.

…

EXAMPLE:

It's in the midst of poetry that I realize my own stark mortification.

If it were divine poetry I would not be so mortified.

…

ABSURD POETRY:

It is in the midst of poetry that I resize my prostitution.

If it were chaste poetry I would lose my license.

Nathan Coppedge

...

TYPICAL

I stole the theory of everything.

I stole a poem.

It had a logical purpose.

It was a poem about everything.

PROGRAMMABLE HEURISTICS AND BIG IDEAS

*HOW TO THINK LIKE A
DIVINE INVENTOR:*

June 11, 2019.

First, see previous:

How To Think Like An Inventor

How to Think like a Super-Inventor

How To be a Smart Innovator

Xeno Invention

Begin by choosing from within one or more extremes.

Make the correct comparison.

Refer to what you desire, which is understanding, using metaphor as necessary.

Additional points if the metaphor is used from the offset, or if the result is useful.

…

Examples of Similar Qualifications:

How To Create Matter (from the Master Permutation)

How To Achieve Nirvana (from the Master Permutation)

Intermediate Omniscience (The universe is an orange)

Micro History of Mind (That cognitive science is lemons)

A formula for happiness (Happiness is a universal orange eaten)

Characteristica Universalis (That language happens simultaneously)

Eta Nothing Theory (That humanity is nothing)

Socrates, On Ethics (Justice has the good life)

PROGRAMMABLE HEURISTICS AND BIG IDEAS

HOW TO THINK LIKE A
SUPER INVENTOR:

First, see previous: <u>How To Think Like An Inventor</u>

1 <u>Learn How to Invent Quickly</u>

2 <u>Make One to Four Great Inventions</u>

3 <u>Pass High Qualification</u>

GENERAL FORMULA:

1. The best.
2. The first.
3. Sole basis for a divine work.

EX 1

<u>Perpetual Motion Flying Machines</u>

- Unique.
- Antiquarian.
- A wonder to behold!

EX 2

<u>Perpetual Motion Q +A</u>

- Gives information.
- Introduces perpetual motion.
- Implicates the future.

EX 3

Souls of Inventing

- They are the right answer.
- They have an early occurence.
- They resulted in the Master Permutation.

4 Make Two Best Inventions

...

PROGRAMMABLE HEURISTICS AND BIG IDEAS

HOW TO THINK LIKE AN INVENTOR:

June 11, 2019.

1. Parts of an Invention

    How to make an invention / How to invent (?)

    WORKSHEET FOR CONVENTIONAL INVENTIONS / BASICS OF HOW TO DESCRIBE A PATENT.

1. (N-dimensional)
2. Rotations, casementing.
3. Functions (boundary, fuel, point, corner, axis, secondary axis, labeled category, labeled axis, substrate, gate, I/O, activation, polarization, activator, dial, instrumentation (visual reader, sensor / detector, 2-way emergency switches), processor, storage, chemical activator, function reader, physical mechanism, heated object, specific I/ O (for example, air), I/O integration, decoration).

2. Lalu Inventors IQ

INVENTOR'S MASTER PERMUTATION, SYMBOLIC VERSION

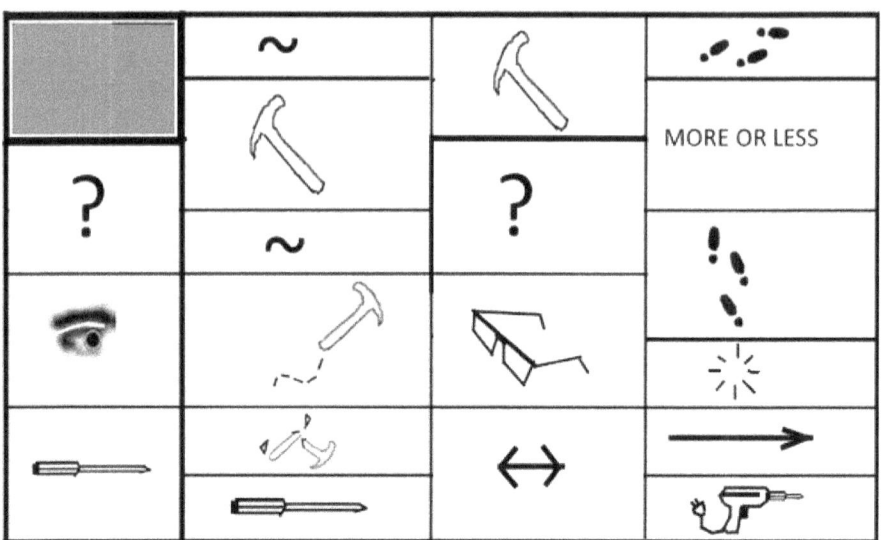

Simplification of 256 categories of ingenious inventions by Nathan Coppedge

3 . <u>Premier Intellectual Dialectic</u>

   1. Take a good example, and create several versions of it.

   Example:

   My primary treatment is to translate or expand Ockham to include philosophical razors designed to provide standards for how to do philosophy: <u>Book of Razors</u>

2. With your experience, extract a higher principle from the original example.

Example:

PROGRAMMABLE HEURISTICS AND BIG IDEAS

One perspective perhaps related to this is that Ockham can also have a 'higher translation' in terms of logical or mechanical (etc.) efficiency.

3. Greatly improve the higher principle by adding another factor, for example, simply doubling it.

Example:

We can then use the principle of efficiency ingeniously to arrive at exponential efficiency.

4. Now, use the improved higher standard as a platform for a body of very new ideas.

Example:

Exponential efficiency can then be used as a platform concept for masterful fulfillment of the logical and mechanical criteria.

5. Find the best general categories within the new system / platform.

Example:

This leads to the general concepts of preferred knowledge and continuous motion machines,

6. Now translate the general categories using your understanding of the general and specific meaning.

Example:

...equals objective knowledge and perpetual motion.

...

4 . 8 New Types of Innovation

1. Alternative to a material or technique.
2. Alternative to innovating through science or philosophy.
3. Consumer alternative to computation.
4. Alternative to missing categories of invention, missing functions, simplicity, or complexity.
5. Major improvements not related to manufacture, design, function, thought.
6. Employers express something new that is not a material need.
7. Alternative to inventing that possibly fits a social need.
8. As far as solutions, alternative to orchestration.

...

PROGRAMMABLE HEURISTICS AND BIG IDEAS

XENO INVENTION:

> Sometimes perpetual motion is survivalism only. Sometimes dunking donuts is diplomacy only.
>
> <u>Are scientific formulas discovered or invented?</u>

Formula:

Possible function, (often different) limited case.

Typical case: paperclips are aesthetical, used only for lockpicking and tying hair.

DIVINE FUN:

June 26, 2019.

You are a human, they are a god. You are putting all your talents into one person. But you become them.

And you have a choice of who to become. In a way you are more free than them. You are the definition of freedom. You created your own future. Everything you were able to create csme to be.

Then enjoy the moment. Quantities of qualities.

PROGRAMMABLE HEURISTICS AND BIG IDEAS

CRAFTING /
THE FOUR SECRETS:

[handwritten notes: The Four Secrets — additional, transigent, dimensional, exceptionary]

1. Additional.
2. Transigent.
3. Dimensional.
4. Exceptionary.

...

Nathan Coppedge

Symbol of Crafting

…

Formula for Crafting:

1. **Additional.**
2. **Dimensional.**
3. **Exceptionary.**
4. **Small transigent.**
5. **Additional.**

…

PROGRAMMABLE HEURISTICS AND BIG IDEAS

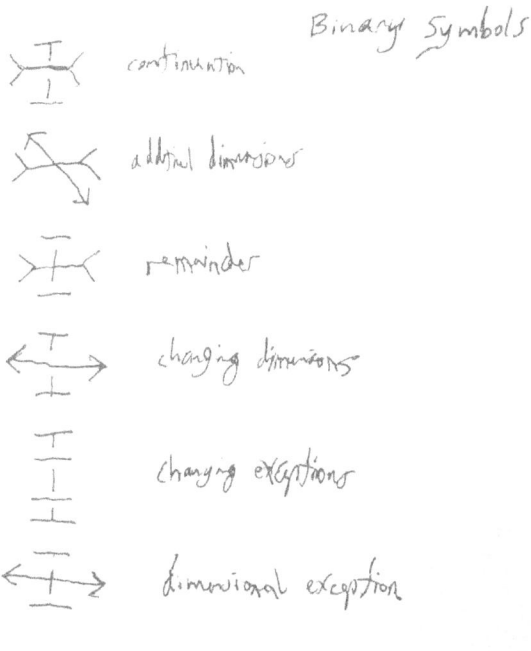

Binary Symbols:

1. Continuation.
2. Additional dimensions.
3. Remainder.
4. Changing dimensions.
5. Chsnging exceptions.
6. Dimensional exception.
   ...

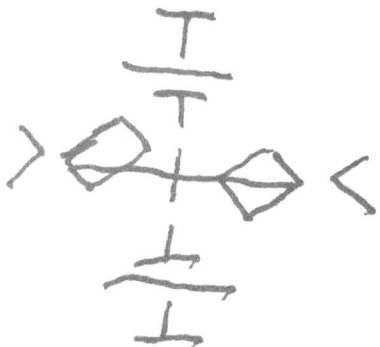

Symbol of the universe.

...

CRAFTING APHORISMS:

There is no hyper-semantics.

PROGRAMMABLE HEURISTICS AND BIG IDEAS

CLARITY:

Actually, on second thought it is possible for some children, It usually involves chimes and large rooms with low cielings and very comfortable seating, or an interesting house with three floors and chimes.

Cold weather and chimes sometimes does it for adults.

As for an enlightened state however, it is normally impossible.

Nathan Coppedge

BREAKING THE 5TH BARRIER:

I go in for a kind of 'access theory'. In access theory of metaphysics, it is a person's behaviors, traditionally called perceptions, which can permit access to lower or higher dimensions. In our current dimension we are experiencing a combination of the 3rd dimension (the dimension of blocks of concrete), and the 4th dimension (the dimension where, according to our feeble senses, multiple dimensions of time might emerge). I call this the 3.5 dimension.

If we consider raw variables, there are many 'dimensions' even within the 3.5. For example, substance, energy, change, and texture might be dimensions. But for some reason these are not normally what mathematicians mean by dimensions. Dimensions to a mathematician can incorporate multiple potentials, hence the difficulty in moving from one to another. Each dimension is arguably much more complex than the previous (an entire 'new dimension'!). Like insects, we appear to live at the level of our education, and learn to live in higher dimensions only when we're completely prepared.

Typologically or categorically, the third di-

PROGRAMMABLE HEURISTICS AND BIG IDEAS

mension is where material meaning arises, and thus moving beyond it may involve significance that is less materialistic, or perhaps even a post-significance concept.

It may help to picture that higher dimensions also have animals, which we are currently competing with in the lower dimension of 3.5. There may be varieties of worms, for instance, that are more adapted to time travel than we are. Whatever it is may typically involve travel beyond materialism. Thus (I know this sounds like God had a bad day) but one possibility is that the fourth or fifth dimension exists as information, or virtual reality.

Although none of these answers are precisely satisfying, access to time travel would explain the barrier to reaching the fourth dimension.

In that case, there are at least four dimensions of space, three being normal spatial dimensions, and the fourth being time, which goes in multiple directions. The fifth dimension may then involve additional permanence, manifestation, or intensity within those dimensions, or it may involve having volitional control over the appearance of all objects in the world.

Don't be shy to compare higher dimensions to Godhood. After all, placing limitations on what dimensions signify may ultimately prove to be a shortcoming.

In mystic traditions it might be said that there is only one 'absolute' dimension, which goes beyond the dimensions we are usually aware of. And then they take back their words, and say it isn't a dimension at all. it is something more permanent.

As for me, I think there might be infinite dimensions of space. Perhaps life as we know it now is only like a leaf floating on the surface of a river!

PROGRAMMABLE HEURISTICS AND BIG IDEAS

ANTIGRAY: 'Imagre, add water, rub upwards'
is like how to prevent gray hair
(from <u>Miscellaneous Spells</u>).

MAGIC AIR CONDITIONING SPELL

Consider: **1. Man's God Solver:**

Ask yourself, you have dreamed many times of the temple's *what*?

For example, you say substance.

Then what do you think about this?

Not yet perhaps.

Then the temple is not yet.

2. Woman's God Solver

Woman is a liar. Name me true and you will be forever.

Being is crying.

Not quite.

Not quite not quite not quite…

Can't Quit.

Quoits, clumps…

Continental Quotes…
….

3. Focus one's powers on cooling.

PROGRAMMABLE HEURISTICS AND BIG IDEAS

MASTER SYSTEMS:

IMPROV: Improvised Thinking, Research. <u>Premier Intellectual Dialectic</u>

1. Improve.
2. Duplicate.
3. Generalize.
4. Interpret.

EVOLVE: AutoComplete. <u>Phenomenal Science</u>

1. Prediction.
2. Gather research.
3. Apply prediction in light of research.

SAVEGAME: Compose: SaveGame. <u>Logos: Rules of the Common</u>

1. Modify.
2. Wait.
3. Codify.

FF: Fast Forward, Noos, Mind.

1. Speed up /process.
2. Stop to look.
3. Engage.

MASTER SYSTEMS /

PREMIER INTELLECTUAL DIALECTIC:

Sept 10, 2018. ("10 grains of salt day").

Other ideas: How can we advance the [field/ topic/ industry]?

Expansion: Guaranteed Foundational Disciplines

OVERVIEW

Improvise: Research.

1. Improve.
2. Duplicate.
3. Generalize.
4. Interpret.

…

MAIN TEXT

1. Take a good example, and create several versions of it.

Example:

My primary treatment is to translate or expand Ockham to include philosophical razors designed to provide standards for how to do philosophy: Book of Razors

2. With your experience, extract a higher principle from the original example.

PROGRAMMABLE HEURISTICS AND BIG IDEAS

Example:

One perspective perhaps related to this is that Ockham can also have a 'higher translation' in terms of logical or mechanical (etc.) efficiency.

3. Greatly improve the higher principle by adding another factor, for example, simply doubling it.

Example:

We can then use the principle of efficiency ingeniously to arrive at exponential efficiency.

4. Now, use the improved higher standard as a platform for a body of very new ideas.

Example:

Exponential efficiency can then be used as a platform concept for masterful fulfillment of the logical and mechanical criteria.

5. Find the best general categories within the new system / platform.

Example:

This leads to the general concepts of preferred knowledge and continuous motion machines,

6. Now translate the general categories using your understanding of the general and

specific meaning.

Example:

...equals objective knowledge and perpetual motion.

...

Ad-Lib Version:

My primary treatment is to translate or expand **area of X** *to include* **quality1.... Xs** *(new combination) designed to provide a way to do* **noun of quality1.**

One perhaps related (perspective) is that **ExofX** *can also have a 'higher translation' in terms of* **property 1, 2...** *(such as logical, mechanical, etc.) ...* **QualityofexampleofX** *(such as efficiency).*

We can then use the principle of the **QualityofexampleofX** *(ingeniously) to arrive at* **AdditionalQualifyingAttribute ... QualityofexampleofX.**

AdditionalQualifyingAttribute ... QualityofexampleofX can then be used as a platform concept for masterful fulfillment of the **property 1, 2** *criteria.*

This leads to the general concepts of basic **[AdditionalQualifyingAttribute ... Quality**

PROGRAMMABLE HEURISTICS AND BIG IDEAS

ofexampleofX for property] **1, 2, etc.** *(such as a sketch of property 1, 2 of exponential efficiency).*

Improve result*(s)* to make it more legible. ... *(for ex, property 2 of exponential efficiency is really perpetual motion machines).*

...

SHORTER VERSION:

[1] Prolific. Ability and willingness to do work which might be significant.

[2] Formal. Evidence of logic, generalization, or physics.

[3] Paradigmatic. Interpreting an important formula or philosophical discovery into something with wider significance.

[4] Applications. 'X is a masterful platform concept because Y,Z may use it'.

[5] Clarify. Improve the wording of concepts Y ... Z etc.

[6] Meaning. Now choose exactly the right words.

...

MASTER SYSTEMS /

THE PHENOMENAL SCIENCE

Sept 25, 2018

OVERVIEW

Evolve: AutoComplete.

1. Prediction.
2. Gather research.
3. Apply prediction in light of research.

…

I. Basic Evolution

Someone does something very basic. They might feel shitty.

They know they are alone and probably no one else will be with them at first. What is new is one single difference: a new invention. It might be a tree thing, or a road thing.

That is what is evolution.

…

II. Advanced Evolution

Completing Trees:

A phenomenal scientific method provides

paths for evolution by immediately completing trees.

> The phenomenal method can be interpreted perfectly through a positive conception or implementation of what exists.

—Perfect Science

III. A Better Sense of Evolution

Oh, the times!

We're healing ourselves!

Possibility is maximized!

A paradigmatic journey!

…

A Preferred Form of Evolution:

1. Implementation.
2. Reaction-completion.
3. Reality.
4. Details.
5. Elaboration.
6. Semantics.
7. Exponents.
8. Confidence.
9. Principle.
10. Nonchalance.

A posteriori:

Ideal Science

...

Perfect Science Based on the Phenomenal Science:

Goes up and down the structure of phenomenal science, understanding it and translating it into scientific statements.

...

Alternate Descriptions:

An alternate science was based on this type of concept of evolution:

1. To have a (complex) motive. (Complexity).
2. To have a paradigmatic structure. (Cause).
3. To have a case. (Conditions such as luxuries or tools).
4. To evolve means simply to exist as a product of age-old time. (Wisdom).
5. Evolution paradigm. (Consciousness).
6. Individual motive. (Application of consciousness to an object).
7. Dynamic state. (Test of applied consciousness).
8. Paradigm of survival. (Elite existence

equals existence at all, survived every time by definition or imply limited resources).
9. Additional thing: meaning. (What you say happens, happens, to a degree).
10. Everything is perspective qualified by degree. (With sufficient power, perception is a sufficient requisite for evolution).

Translated:

1. Resources.
2. Organization (skip #3 above).
3. Efficiency.
4. Realization.
5. Testing.
6. Advanced testing.
7. Obviousness.
8. Significance.
9. Power.

Further Minor Perfect Science Methods:

1. Based on existing evidence? 2. Provide media attention. 3. Add to list of collaborators. 4. List experiments with total list. 5. Find statistics.

...

Notes:

Infamy if it compromises, becomes immune. Immunity if it is timeless has no problems. Or humans are more evolved than four other alien species.

(—What are some of the most interesting explanations for the Fermi Paradox?)

Alien Genders 1: Human genders: Substantial Theory of Genders

Alien Genders 2: Compromisers: Not enough credit versus funny credit (after basic, sbstract), ambitious and flitty, Ambiguous and Supercilious. Realization and Manifestation.

PROGRAMMABLE HEURISTICS AND BIG IDEAS

MASTER SYSTEMS /

SAVE GAME

(LOGOS: RULES OF THE COMMON):

It has a form.

It holds to the form.

People react to the form.

You can modify the form.

And the form means something.

Edit and wait

—*Philosopher on a plate.*—

…

OVERVIEW:

Compose: SaveGame.

1. Modify.
2. Wait.
3. Codify.

111

MASTER SYSTEMS /

FLASH FORWARD

FF: Fast Forward, Noos, Mind.

1. Speed up /process.
2. Stop to look.
3. Engage.

PROGRAMMABLE HEURISTICS AND BIG IDEAS

MASTER LOGICS

NOT TAKING DAMAGE

'I could develop that area'.

Lie if the area is too dangerous.

MASTER LOGICS

SOUL OF ENERGY: THE UNIVERSAL DEDUCTION OF NATURE

1

Positive principle:

Absence is always a contrast.

2

As-is.

Energy is the realization of anything.

X is Important?

For ex, They want to drain the Sun?

Just a sign of massive amounts of energy.

(Everything we know has involved the sun except perhaps antimatter).

3

Everything is inevitably *potential*.

If it could become *this*, it coild also become *that*.

PROGRAMMABLE HEURISTICS AND BIG IDEAS

Or, it was *inevitable*.

Otherwise it was incapable of the first thing.

So, potential is natural.

4

What is natural has nature.

Nature is the way of things.

The way of things confirms how things are.

MASTER LOGICS /

AN IMPROVEMENT ON
ARISTOTLE'S SYLLOGISMS:

RECORDED AUG 24, 2018.

Coherently, syllogisms are mostly reducible to a number of forms, here ranked subjectively from most difficult to most easy. Note that these are NOT categorical deductions in the coherent sense, but are rather CAUSAL forms of inference.

—(NOTE: 'In common with' categorical deduction, this assumes A and B are primary and opposites oppose across the diagonal)—

(1) DBCA

Conclusion D follows from the result of A, concluding that the result of B can be infered from A.

(2) DBAC

Conclusion D follows from the result of A, concluding that the result of B follows from A.

(3) BDCA

The result of A is conclusion D, because the result of B is what we already assumed (A).

PROGRAMMABLE HEURISTICS AND BIG IDEAS

(4) BDAC

The result of A is conclusion D because the result of what follows A is what precedes D.

(5) CADB

What follows from B implies what we assume (A), because our conclusion implies what follows from A.

(6) CABD

What follows from B implies what we assume (A), because what follows from A leads to our conclusion D.

(7) ACDB

With our premise A we find what follows B, for B is hidden in our conclusion, so what follows B will follow B, as B has occurred.

(8) ACBD

With our premise A we find what follows B, for our conclusion follows from B.

MEMORY, PROCESSION, TELEMETRY, CONDITION, PROCESS, PARTY, POSIT, EXIT. (2018/10/10)

SIMPLE VERSION IN REVERSE ORDER:

Goodbye, That's true, I'm with you, Something happened, Maybe, I see you, Follow me, I'm thinking.

VARIATION BASED ON TOP FIVE CATEGORIES OF HUMAN INVENTION:

Let's try it right?, It works, Let's pretend!, It has an unconscious quality…, Something big! --<u>What do you think are the top 5 inventions that have greatly improved the world?</u>

…

Some credit due to <u>YYang</u> as early as 1995.

For traditional syllogisms of the Aristotlian tradition, see: **<u>Causal Inference</u>**

See also:

<u>What is an example of a causal inference?</u>

<u>Accompaniment for Causal Inference</u>

<u>Categorical Studies</u>

NOTE: A similar logic may be found for example, in something similar to the Lycian Lecture:

> So, we consider problems.
>
> We also consider solutions to them.

PROGRAMMABLE HEURISTICS AND BIG IDEAS

> We consider principles as well.
>
> And general problems and solutions.
>
> And advanced things.
>
> And difficult problems.
>
> And special problems.
>
> And general systems.
>
> —<u>Study of Ends by Nathan Coppedge</u>

This writing seems to describe two basic categories of 'ends':

(1) Problems, and (2) Advanced Things.

These are further subdivided into 'problems' solved by 'advanced things':

(1) Problematic solutions are solved by forming a standard of difficult problems.

(2) Problematic principles are solved by enjoyable problems.

(3) Problematic general problems and solutions are solved by general systems.

…

[A COMPATIBLE SET OF ARGUMENTS:]

(1) The Argument from Personal Benefit.

(2) The Argument from Useful Benefit.

(3) The Argument from Sufficient Benefit.

(4) The Argument from Inevitability.

(5) The Argument from Description of Existence.

(6) The Argument from Benefit to the Universe.

(7) The Argument from Logical Completeness.

(8) The Argument from Overbearing Relevance.

—How can you prove a philosophical proof?

(This writing may have contributed to the theory: Critique of Longinus, On The Sublime)

…

PROGRAMMABLE HEURISTICS AND BIG IDEAS

This may help with the deductions, a kind of coherent arrangement of simpler logic statements:

1. Everything = Everything
2. Everything = Something.
3. ==
4. Some Definition is Everything.
5. Something Not= Everything.
6. Some Definition Not = Everything.
7. Every Definition Not= Coherent.
8. Law of Identity.
9. Possibility Not=
10. Something is not something.
11. Nothing is nothing.
12. Definitions are never complete.
13. Everything Not= Every Definition.
14. Not Everything is a Definition.
15. Everything Not Always Something by Some Definition.
16. Everything Sometimes Has a Definition.
 —Nathan Coppedge, Nov 9, 2018

—The Essential Proof

MASTER LOGICS /

METHOD OF TRADITIONAL INDUCTIVE INFERENCE /
 "INSIGHTFUL INFERENCE":

You need a pass code to do mathematics?

Well, yeah, sort of.

But you also need a pass code to be a Greek philosopher!

Rats.

But probably it doesn't require a pass code to be a Greek philosopher, that's what we're thinking.

And we're probably right.

But equally, you could be completely wrong.

But that's a bad idea.

PROGRAMMABLE HEURISTICS AND BIG IDEAS

MASTER LOGICS

X-ICAL DEDUCTION:

It's a good principle that nothing is a X.

Nothing is a X.

…

Function X.

If good, good function.

MASTER LOGICS /
MASTER DEDUCTION:

Process Y of what opposes X. Proves:

Opp of Y applies to X.

…

And you have to phrase it properly.

1. In the first phrase mentioned second use 'It is...' or 'We have' etc.
2. Qualify terms.
3. In the second phrase (coming first, added later) make double use of the first major term in this part.

…

EXAMPLES:

1

We have to teach subjectivity if we teach traditional religion.

Objectifying the soul makes self-love look useful.

…

PROGRAMMABLE HEURISTICS AND BIG IDEAS

2

There is a natural limit to processing the imagination.

Automatically processing confounds the imagination.

...

3

It produces insanity to utterly lack meaning.

What is reasonable is not without meaning.

...

It is not common for immortals to die.

It is common to live a mortal life.

...

This idea may be due to **Katy Ruben**

MASTER LOGICS / THE HARD LOGIC:

1

EXAMPLE:

I'm not going to sell my soul for pain.

.: I'm not pain.

GENERAL FORM:

I'm not going to have a problem with P.

Therefore if Q is the only solution to P, Q.

Modify to account for alternate Q.

Use alternate Q to predict alternate P.

2

    X (hard).

    Y (easy).

    Something makes X easier.

    Something X seems Y.

    Y is now similar to X, because

hard is like easy.

Someone makes Y a successful X, so hard is sometimes the definition of easy.

(Now) Y seems X. For example, a fool might be a savant.

Someone makes X a successful Y. Some fool might become the greatest savant.

Now Y and X are simply paradigms X-Y: Y-X, because X is now yet another exsmple of continuum X, as is X of Y.

—Prediction Tools 2019-02-19

…

(Originally September 14, 2018).

MASTER LOGICS /

THE OLD LOGIC:

It's absolutely absurd to X.

But that is because the opposite was true before.

PROGRAMMABLE HEURISTICS AND BIG IDEAS

SECONDARY SYSTEMS /

COINCIDENTAL LOGIC:

If someone is balding but they take Jiaogulan and have an infantine ability to grow their head, they might avoid the condition of balding.

This might not apply to everyone, but hypothetically at least it might work, at least for a time.

The coincidental logic is, was the problem of the balding man arranged to clue people in to this solution to balding?

Even if it was not, it might be used coincidentally as if it was!

Hence, coincidental logic.

SECONDARY SYSTEMS /

MAGIC SOLVER /

MAGIC PROBLEM-SOLVING:

> If one wishes to culture the soul, one may also need to culture the intelligence!

[Coherent Systems 2.B.1.A.4.]

XQ

Magical train (X), magical gain (Q).

They are the same!

Is X magical with Q? Magical Q!

Helps if Magical X.

PROGRAMMABLE HEURISTICS AND BIG IDEAS

SECONDARY SYSTEMS /

FORMULA FOR INTUITION:

This is meant as a very crude example:

[Some] wombats [don't] fly.

…

Qualification… something… reasoning.

[No] building is [ambiguous] when it falls apart.

[Everything] human [arbitrates] the animals.

[The] rule is the [universal exception]

…

1. Some… don't.
2. No… ambiguous.
3. Everything… arbitrates.
4. The… universal exception.

…

Some credit due to <u>YYang</u> and <u>Katy Ruben</u>

SECONDARY SYSTEMS /

PRIVATE LANGUAGE:

One makes an impression on one's inner life-world, like the sound:

Cloem.

Ordinarily one might think, maybe it is an attempt to express the word 'clue' or perhaps 'quarem' or some love for an external object.

However, this time the participant just says: "It's a private thought".

The point is, private language is communicated differently.

…

…

It's an internal feeling, no point in communicating it, it is not a communication.

Some insight from JF .

PROGRAMMABLE HEURISTICS AND BIG IDEAS

SECONDARY SYSTEMS /

FORMULA FOR 1ST AND 2ND WORLD PROBLEMS:

Appreciating... perpetual motion.

Wanting... to be a writer.

...

Formula:

Indulgent stretching... something clearly luxurious.

2nd World Formula:

Indulgent stretching... something not luxurious in the first world.

SECONDARY SYSTEMS /

FORMULA FOR INVENTING SOMETHING:

EXAMPLE

> I always sleep for eight hours every night!

> Not anymore, that's your calling sign! That would have been an excellent joke in Ancient China!

> Irreverent sense of humor! You must have invented it!

(Prior information)

Other contradicts information.

Call up prior information. Perform a call (function) on the information.

Other person is at wits end to refute the call function, as they have expressed non-commitment to the basis for the call for the first time.

The function that is called becomes the original invention.

PROGRAMMABLE HEURISTICS AND BIG IDEAS

SECONDARY SYSTEMS /

FORMULA FOR FANTASY:

Possibly very delayed writing.

…

Much more functional, much less practical.

Nathan Coppedge

SECONDARY SYSTEMS /
FORMULAS FOR THE GOOD:

May 2, 2018.

FORMULA

The rational emotion which gives one confidence.

OR, Abstract, perfect relief.

PROGRAMMABLE HEURISTICS AND BIG IDEAS

…

1. Language, logic, wisdom, magic, & meaning.
2. Perpetual motion, robotics, physics, time-travel & computers.
3. Money, clothing, secrets, ethics & laws.
4. Architecture, vehicles, plumbing, interfaces, & memory.
5. Immortality, evolution, medicine, beauty & grooming.
   —<u>What do you think are the top 5 inventions that have greatly improved the world?</u>

…

Here's a possible list of the Top 10 greatest inventions:

1. Thought / philosophy.
2. Perpetual motion machines / paradise.
3. Immortality / godhood.
4. Calculus / intelligence.
5. Logic / analysis.
6. Electricity / appliances.
7. Computers / internet.
8. The car / The wheel.
9. Medicine / chemistry.
10. Materials science / Physics.

…

Alternate description:

1. Good.
2. Good location.
3. Super good.
4. Something to do when you're good.
5. The natural result of doing something good.
6. Good tools.
7. Super good tools.
8. Super good location.
9. How to be good if you're not good.
10. Hope to be good.

…

Another alternate description:

1. Invention.
2. The Greatest Invention.
3. Someone Great.
4. Genius.
5. Genius Thing.
6. Genius Applications of Things.
7. Extension of the Genius Application of Things.
8. Prior Art.
9. Instructions for Someone Great.
10. A User Manual.

PROGRAMMABLE HEURISTICS AND BIG IDEAS

SECONDARY SYSTEMS /
VARIABLE ANALYSIS:

If X, would it have changed in a similar manner?

What about 'root X'?

What about the root 'situation of X'?

What about the full 'extension of X'?

What about the 'interpretation' of X?

What about the 'selection' of X?

What about the 'physics' of X?

SECONDARY SYSTEMS /

SEARCH ALGORITHM:

Google (google). Yes. Good sample. Something good. Something prime. Other thing-better. View of the world. Place of power. Preferences choice. Identical… feather.

PROGRAMMABLE HEURISTICS AND BIG IDEAS

USEFULNESS OF LACK OF INFORMATION:

1

A person has a terrible gambling problem. But he does not know there is a casino on the next block.

As long as he doesn't know there is a casino, he will have less of a problem.

2

Charlotte is a nymphomaniac who has 266 venereal diseases. If she knew John were straight, she would try to sleep with him.

She doesn't know he's straight, so he doesn't get venereal diseases.

3

Someone without U.S. citizenship develops a cure for cancer in the U.S.

If he was caught not being a U.S citizen, in this case he wouldn't develop a cure for cancer.

4

The first female President stays secret about her sadism with her husband.

She turns out to make a good president, because she inspires women around the world. Thus her sadism problem is a private concern.

(By the way, I'm not a masochist, I just thought this was a good example where the ends justify the means).

NEGATIVITY STUDIES

Eileen called me 'skinny skeleton' when she meant 'skinny apple'. Although she wanted to call me skinny apple, in this culture the word skeleton was seen as more appropriate…

--Childhood experiences

NEGATIVITY STUDIES /

FORMULA FOR EPITAPHS

July 19, 2019.

Be it that X (died, etc rhyme as necessary).

The / it's Y continues on.

Opp Y still is so (for ex, nonetheless, here he lies).

Such was X in Xness, that it became the opposite.

PROGRAMMABLE HEURISTICS AND BIG IDEAS

NEGATIVITY STUDIES /

PROOF OF THE UNDESIRABLE:

For Ex,

- Self-love.
- Death.

We don't want equality.

We don't want pain. And,

We don't want death.

NEGATIVITY STUDIES / ARGUMENT FROM BROKEN REASON:

Because it would feel better than having a broken head.

If I had a broken head is the only reason I'd agree to be a sh**head.

That's absolutely the reason.

That's what I said.

Would it be more humble to have more reason?

Would it be more wise to be less humble?

(What if words matter? What if words don"t matter? What do I have to lose?)

They seem to play a different sleazy system for everyone else.

PROGRAMMABLE HEURISTICS AND BIG IDEAS

NEGATIVITY STUDIES /

LOGIC OF EVIL:

June 22, 2019.

1. [Anything but the answer]
2. [Caring can be the cruelest thing, if it's a cruel world in which there is little if any suffering].
3. [That is what shouldn't look evil but does, for example in Babel, conscious intentions].

…

Soullessness: What about from another angle? If there is no other angle, that suggests an element of soullessness.

For example, gothic architecture in a map. Flat earth. Not being philosophical. Not having senses.

…

The paranoid, irrational conscience taking risks.

(irrational == law == risk)

OR,

Manifesting deliberate frustrations.

(frustration <-> real = important).

NEGATIVITY STUDIES / LOGIC OF DEATH:

Everything about death is an explanation of this:
"A Chinese Sorcerer died for immortality."

For Example,

In some emotional sense (he was an emotional person)…

True death is immortal.

To be great is to die.

Immortality requires the correct medicine.

They were things he may have said, or Guo would have put them into his mouth.

Great is the God, was Guo's fishy excuse.

You should know better.

Not a good piece of work.

It all sounds so fine if you're secretly nobody.

If you're anyone, anyone except the one who died.

PROGRAMMABLE HEURISTICS AND BIG IDEAS

NEGATIVITY STUDIES /
FOOLISHNESS / HARD PROBLEM OF RA-
TIONALITY:

Vis. Hard Problem of Rationality.

It is the problem that says that laws are absurd.

Someone (it wasn't me) said all of rationality amounts to basic common sense decision-making.

It's like you could try a little extra flourish at that point.

But at that point it seems to get a bit iffy.

Like it requires a major enchantment.

The hard problem of rationality: it makes people into something like f**ls.

It is the problem educators avoid because it essentially teaches the opposite of anything one might teach.

It is foolishness in other words.

It might win, but probably not. It is total tripe unless you're right.

Ancient Balogna tries to look lucky.

NEGATIVITY STUDIES /

ANIMALISM / WHAT CAUSES ONE TO BECOME AN ANIMAL:

To be a devil, but not to know.

EXAMPLES:

If Kwang Kuo (Zheng Guo) becomes a Bird of Peace.

If Nathan Coppedge (Asceticurus) becomes a lemming.

PROGRAMMABLE HEURISTICS AND BIG IDEAS

NEGATIVITY STUDIES /

FORMULA FOR GHOSTS:

1

Guo thought the Tall One died.

Or talk to the Sorcerer.

2

Add probability.

Nathan Coppedge

NEGATIVITY STUDIES /

FALSEHOODS:

Nov 15, 2018:

Falsehood is simply the opposite of anything true.

If you have a method for devising truth, the opposite of a given result is false in regards to the same query.

PROGRAMMABLE HEURISTICS AND BIG IDEAS

NEGATIVITY STUDIES /

UNFAIRNESS

Jan 27, 2019:

Basically,
If someone wanted to be unfair, they could.

…

1. If they are stupid, they can trick themselves and believe anything.
2. If they are average, they don't have to feel guilty about anything.
3. If they are smart they can make a diabolical argument.
4. If they are genius they can have the laws of reality that they prefer.

NEGATIVITY STUDIES /

DETECTING HISTORICAL GUILT:

1. Are you great?
2. Is there something greater than paradise?

A: Two honest yesses is the worst. An honest Yes dishonest Yes is second worst. A dishonest No, Yes is third worst. An honest Yes, No is the fourth worst. A dishonest Yes, honest no is fifth worst. Two dishonest nos is good, or avoiding the questions.

Apparently evil fits the pattern of some type of indulgence (consumerism, temptation).

PROGRAMMABLE HEURISTICS AND BIG IDEAS

NEGATIVITY STUDIES / ENTROPY /

THE BASIC CALCULUS OF EVENTROPY:

Let's say there is entropy.

Suspicious positive energy might come ftom entropy.

Then positive energy is worth avoiding.

But consider this may be a sign of entropy.

Therefore, hold energy above entropy.

There is an exception: if the energy is evil.

If the energy is evil, be moderate.

If energy is necessary, hold energy above entropy or be moderate.

NEGATIVITY STUDIES / NONSENSE REASONING:

1

I invented psychology once long ago.

Makes sense if I'm not insane.

But I am insane.

There! I just proved I didn't invent psychology!

2

I don't want to sleep, man, that doesn't make sense!

If it made sense I'd be asleep!

Why would it make more sense later?

Maybe if my life makes less sense!

3

Man, I shouldn't say 'man' so much.

Man, oh man.

PROGRAMMABLE HEURISTICS AND BIG IDEAS

NEGATIVITY STUDIES /
(FORMULA FOR) THE FORMAL ZERO

> Nothing does not have any relativity. It is pure nothingness. —
> Ragesh Nair

In the comparison of coherent and other types of systems…

By <u>Categorical Deduction</u> . Neutral, content-less.
- By constants: Not constant.

By <u>Coherent Exceptions</u> . Not varying.
By <u>Naive Realism</u> . Non-trivial.

By <u>Paradoxes</u> . Not solvable, non-solvable, problematic, and not being a solution.

By <u>Irrationality</u> . Not breaking the rules.

By <u>Incoherentism</u> . Not coherent.

By <u>Definition of Neutrality</u> . Not contradictory.

By <u>Informalism</u> . Informal.

By <u>The Undecided Exception</u> . Coherently incoherent.

By <u>Coherence with Relativism</u> . Not preferred, also Singular, given to relativism.

By <u>Nonsense</u>. Does more than negating its opposite, otherwise nonsense or a real system.

By <u>Impossible Logic</u>. Does not adopt a definition of impossibility, is not a superior concept. Trivial.

So, here is what is required to organize the results:

Neither trivial nor non-trivial, not constant or varying, not coherent, yet content-less, not contradictory, solvable, unsolvable, or problematic. Relativistic and informal. Does more than negating its opposite yet is informal, or is nonsense or a real system but is informal. Doesn't break the rules.

—<u>Can logic be wrong? Why?</u>

Clearer statement of some of these results:

<u>The Opposite Argument</u>

PROGRAMMABLE HEURISTICS AND BIG IDEAS

NEGATIVITY STUDIES / FRUSTRATION / CALCULUS OF FRUSTRATION:

Frustration is frustrating, why doesn't X realize that?

If X always realized that, X would always be frustrated.

FRUSTRATION AND COMPUTERS

Frustration occurs when something is trivial and difficult.

Let's choose something that is not frustrating.

It may be something un-trivial and easy.

Otherwise, it will will be something un-trivial and difficult, or something trivial and easy.

Or, just something easy and adequate all the time.

Nathan Coppedge

NEGATIVITY STUDIES /

ANALYZING STUPIDITY:

I do not consider myself stupid (I was in the Top 10 in my semi-urban high school). However, having big ears has forced me to confront many people who assume I'm close to retarded, and this has led me to think more deeply than some about the root causes of stupidity, and how it might never be stupid happiness, or at least it will always seem spiritually justified to feel genuinely happy. As far as I know, stupidity is caused by:

1. Extreme suffering from an early age. Many low-IQ people have extremely painful brain damage. The tranquilizer chemicals that develop in response to physical trauma may 'dumb' the brain, but this is not the somehow 'intentional' or 'unavoidable' dumbness that people expect. It is possible people with brain trauma are better adapted to trauma than those without the damage, and it may have small benefits like reducing sensitivity to a harsh environment or forcing them to try harder to survive without overthinking. In some cases brain trauma may occur because of a martyr complex.
2. In rare cases happiness or happiness at a price. Some people who look stupid to others aren't concerned for their

behavior due to alternate emotional priorities like a will for happiness, or because they took heavy drugs.
3. Seeking advantages and coping with problems that result. Some people that look stupid are in the wrong social circumstances or have tried too hard to fulfill ambitions which are unrealistic given how they feel about themselves. For example, people who are forced to adapt to an unfamiliar social climate.
4. Being overwhelmed due to lack of experience. In many cases someone may simply not have the exactly right exposure or social pressures to succeed in looking smart. This may for example include categories such as people with angry parents, people who have not discovered their niche, people who rebel, and people who are mentally physically or spiritually distracted.

If happiness is hard to buy, it is likely anyone stupidly happy feels justified, but more likely everyone who looks stupid is spiritually distracted, either attempting to enjoy life, or coping with suffering.

There is one small problem with justifying intelligence: not having a choice. Searching for happiness may have few options for people who are not already happy, and so it is likely most people are suffering, and so it is unlikely looking stupidly happy would be less justified than intelligence. But then, no one is really dumb unless you mean that they are struggling to find happiness.

Also, much stupidity is simply hyped by the media.

Stupidity is obvious from the above, but it's not in evidence from an internal angle. In other words, it tends to be someone's selfish means to an end, and that makes it hard to understand.

*NEGATIVITY STUDIES /
MEANING WITHOUT MEANING:*

Based on a dream 2018/10/01:

Let's say someone proves something in a dream…

But that is not meaning in the integrated, meaningful sense.

1. You show someone something.
2. There is improved clarity.
3. You try to improve communication.
4. Other 'facts' enter.
5. The other person communicates something.
6. An arbitrary terminus is reached.

This is much like a (type of) dream which somehow fails to represent anything symbolic.

NEGATIVITY STUDIES /

LIES THAT AREN'T LIES:

(1990)

There are cases where if something is prepared properly, it does not have to be realized before it counts.

Simply holding a standard of a certain quality may in some cases allow evidence to be unnecessary.

Cases include such things as:

- Prior evidence.
- Providing much better intelligence than the evidence provides.
- Knowledge of a hidden agenda.
- Disqualification of evidence (potentially in more than one way).
- Providing a preferred tactic or technique.
- Appealing to sensibility, mentality, level of education, and popularity.
- Convenience / appropriateness for the moment.
- Correlation with another more appropriate data set.
- Parsimony, simplicity, efficiency, ingenuity.
- Awareness of deep natural or philoso-

phical dangers.
- Psychological conduciveness, for example, contributing to happiness or sensoria.
- Agreement with a political or social agenda, easiness, commensurability.
- Situatedness, strategic usefulness, language advantage, knowledge of terrain.
- Ongoing awareness, communication pipeline, diplomatic concern.
- Larger concern, scale of concern, global issues, matters of greater importance.
- Justice, ethical concerns, emotional consideration, understanding personality and culture.
- Rapport with concerned persons, ability to exert influence, hard politics.
- Special concern (ability to seduce, drug, apply additional manipulation), special abilities.
- Dynamic considerations, competence, reliability, specific considerations.
- Operational details, resources, maintenance, secrecy.

NEGATIVITY STUDIES /

FORMULA FOR RELATIVISM;

X and Y are Opposites (for example, Conventional Beliefs and New Ideas).

(X and Y) are Z (for example, 'silly': new is always silly because it doesn't use the evidence, conventional is always silly because it is replaced with the new).

Z is Relative (we know X and Y are 'relatively silly' because they are opposites, and both are silly).

Therefore, X and Y are Relative.

…

PROGRAMMABLE HEURISTICS AND BIG IDEAS

NEGATIVITY STUDIES /
FORMULA FOR PROBLEMS:

> But it had a little honey in the bottom.
>
> —The Soul of Problems

[Coherent Systems 2.B.1.A.2.]

A problem is a problem if there is unexpected resistance, a general requittal of one's influence, a certain incompatibility with one's motives accompanied by a lack of reassuring qualities about the provision for alternate options: a mixture of stubborness, opposition, difficulty, and doubtfulness.

Nathan Coppedge

NEGATIVITY STUDIES /
PROOF OF WORSE THAN PAIN:

I was a baby lying there with pins sticking out all over.

All I had to defend myself was euphoria.

Except that was a big lie.

I was supposed to defend myself with a Big Lie.

But that was my only defense against pain.

So I had no defense.

However, it was worse than pain, because it was all pain and I wasn't supposed to defend myself.

PROGRAMMABLE HEURISTICS AND BIG IDEAS

NEGATIVITY STUDIES /
TRAGEDY:

X [comment on how evil it is]: It looks skeletal.

Well, it is X: Well, it is skeletal.

Sick remark: Let's not turn into skeletons.

Filling the emptiness: You know, you'll feel better.

Sadness.

NEGATIVITY STUDIES /

DESTRUCTION

Time, Contradiction, and slight of hand.

Note: There is an emergent logic sometimes called evolution that there must be infinite logic in infinite destruction, and so, destruction measures what is unworkable. It is expensive, pointless, elaborate, and advanced to create destruction.

PROGRAMMABLE HEURISTICS AND BIG IDEAS

NEGATIVITY STUDIES /

CALCULUS OF JUSTIFIED ANGER:

It allowed me to move my sleeve.

It wasn't wasted.

(X could be wasted).

(How is it relevant that you Y).

NEGATIVITY STUDIES /

SIN

Negativity and dishonesty in an imperfect world.

There is sin if:

1. The world is sinful.
2. If negativity is a sin, and there is negativity.
3. If dishonesty is a sin, and there is dishonesty.
4. If negativity results in concealment of truth.
5. If dishonesty breeds negativity.
6. If all truth is destroyed.
7. If dishonesty and negativity is a vicious cycle.

PROGRAMMABLE HEURISTICS AND BIG IDEAS

NEGATIVITY STUDIES /

HOW TO KILL NAZIS:

Integrate their ass.

Apply scushari and furushiki.

Nathan Coppedge

PROGRAMMABLE HEURISTICS AND BIG IDEAS

OTHER SYSTEMS /

QUANTUM PERCEPTION / BREAKING INTO QUANTUM PERCEPTION:

First, if you want to look smart, be honest.

Then, if you want people to try to understand you, be complex.

Anything else should be a choice, in which you get both or really the opposite of both.

Specialist or observer: the ultimate expert.

Intuition or private insight: something people think is easy to understand.

OTHER SYSTEMS /

DIALECTICAL INFERENCE:

To do this we have to take an example.

For example, snow falling.

If we have special information about the snow, we can say, for example, 'snow makes us slow'. Snow == Slow.

Now we 'approve' the theory.

Let's say, we have always observed snow makes us slow, and it applies to us, it's good enough to be true from our point-of view.

But WHY?

To answer this question we have to work backwards a bit… and forwards.

If 'slow' were wrong — if 'snow' was wrong, then slowing time would be wrong.

But slowing time is NOT wrong, unless there is an additional thing wrong with snow, specifically something wrong with snow being slow.

However, since everything is time, the only thing that could be wrong with slow is if it makes us go FAST.

PROGRAMMABLE HEURISTICS AND BIG IDEAS

Therefore, the problem with snow is that it makes us skip time, or there is something (some other possible thing) wrong with our experience of time such as pain, or there is something good about going fast, or some principle of uncertainty.

And the same goes for any other concept, not just snow, while time holds.

—Concept due originally to Professor Vu, who claimed it was originally an attempt to prove Dedekind's Uncertainty Principle.

Nathan Coppedge

OTHER SYSTEMS /

FORMULAS FOR SUBJECTIVITY:

Aug 17, 2018

1 Chosen location with an elaborate explanation attempting to communicate or not about motivators.

2 Rule Against Intermediates Vs. The Counting Problem.

> Once we add quantum, this might resolve against definite properties that are not observable. —<u>Wheel Problems</u>

3 Train of Subjectivity: At some point you get off the line, [and and and etc.]

4 Different* perspective.

...

Note: *Some credit may be due to <u>JF</u> , Katy Ruben, Nico Banach, and R Rienzi.*

PROGRAMMABLE HEURISTICS AND BIG IDEAS

OTHER SYSTEMS /

FORMULA FOR ABSOLUTENESS:

Sept 9, 2018.

What gives X is absolute if X is absolute.

For example, X could be evidence.

X could even be superior but not perfect evidence (in other words, a relative claim), and still seem absolute situationally.

Now if X is exclusive as far as the absolute, then it provides the only means to the absolute, relatively or not, but that is not necessarily the case.

What is absolute but is not X (is not the precedent for the absolute) must be experienced directly, therefore, the only X which points to the absolute is empirical knowledge, as X must be an experience of the absolute or does not qualify as knowledge.

Otherwise, knowledge is not what is meant by X, and there would be no knowledge of the absolute, except directly, and X and direct knowledge can be the same.

Since we know X is the experience of the absolute, we can say that the only evidence we have of the absolute is scientific evidence, or at least some form of experience.

Nonetheless, since X *supervenes on* the absolute, we must argue this scientific knowledge *is absolute* if we want to prove the *experience of* the absolute is absolute, unless the experience of the absolute is direct. This is true even if the knowledge is not scientific, in which case it is some other experience which is not scientific.

So, there is an additional problem of how to determine absoluteness indirectly. However, insofar as absolute evidence is evidence of the absolute, it must constitute evidence of absoluteness, relatively or not.

Therefore, if it is not relative, it is absolute at least in the sense of not being relative.

If something is not observed to be relative, then, it is absolute evidence.

Qua evidence there will be no better definition of the absolute than what is considered to be evidence of the absolute, and so the best evidence of what is not relative must be admitted as evidence, or there is no absolute measure of the existence of evidence.

Therefore, with a relative measure or not, evidence is merely the best evidence of the absolute. And so, the relative measures the absolute. And so, a partial conservative definition of the absolute in lieu of further criteria is relativism is measurement.

PROGRAMMABLE HEURISTICS AND BIG IDEAS

Relativism == Measurement

Absolutes must be measured, so…

QualifiedAbsoluteness == Relative Relativism

(A Qualified Standard of Measure).

OTHER SYSTEMS /

FORMULA FOR EXCEPTIONS:

A single point or contingency within an unobserved dimension.

Variation in variation.

Justified or arbitrary alternation.

Reducing to zero and extending what replaces the zero to infinity.

EXAMPLES:

1

> Concept you hadn't heard of before, really exists.

2

> Concept you didn't recognize, doesn't exist.

3

> Another thing, classed as miscellaneous.

4

A number of things, undefined relation.

5

Application, requires new method.

6

Surprising result, after work.

7

It, does something unexplained.

8

Theory, aberration.

OTHER SYSTEMS /

FORMULA FOR MYSTERY:

Primary Formula for Mystery: Life is a conundrum of esoterica.

Details:

It means several things:

1. Life has a great deal of technical meaning.
2. We don't want to solve every problem, because then we'd be dead (some problems aren't for solving, because they aren't really problems, but rather hesitant guesses about something that will arrive in the future, not necessarily death).
3. Life is an opportunity for discovering solutions to the problems that we stand to benefit by if they are solved.
4. There may be some things in life that are unexpected, almost occult, because life emerges from so many mysteries, and not all of the mysteries remain mysteries.

Therefore there is a conundrum, → Therefore there is a mystery → Therefore life is esoteric.

PROGRAMMABLE HEURISTICS AND BIG IDEAS

OTHER SYSTEMS /

CHILD LOGIC:

Everything seems sticky and gummy, therefore it doesn't matter.

It was an accident heaven saves.

It is (adult concept) so you might be wrong.

You might be wrong (word of affection).

Are you (big negative term) (loving term)?

What is so (evil, difficult term)?

But (why, you are so stupid) (daddy, mommy)?

What if the (ridiculous to an adult) monsters get me?

Will it (something sad and wonderful) tomorrow?

Are you (term child doesn't quite understand) today?

What is (the most, biggest) (adult concept they have just learned to use)?

Why is it so (strangely good / bad)?

OTHER SYSTEMS /

TOO LOGICAL LOGIC:

May 16, 2019.

We need to be in hell, because it's the good place. We need our ability to argue!

Now, we might need our problem-solving skills!

What luck, what luck is there at all? We have done it!

There is a perilous way out of the labyrinth!

Blocked by a single fateful door!

Oh my God! What have we done?

This is an opportunity that is opening up!

So far we are swallowed past the door, which opens into intimations of expressive dimensions beyond.

PROGRAMMABLE HEURISTICS AND BIG IDEAS

OTHER SYSTEMS /

RULES OF DESIGN:

Formerly called Points of Metaphysical Artistry...

Arbitrarily could also be called Nature Magic Part 1.

1.

- To have a continuum of dimensions, and means of changing the world through progression or aesthetic choice.
- To develop each dimension in the metaphysical continuum for its own sake: physically, mathematically, inventively, as a form of artistic variation, as a form of information continuity, and practically.
- Using knowledge as a 'second source' of nature: permitting beneficial spells, fantasy powers, phenomenology, idea-lese, apperception.
- To turn ordinary things like machines, computer code, folding fans, and seduction into an ordinary and safe part of nature.

2.

- To have perfect standards, and use divine art to benefit those who are not covered by the existing standards. To make disappointed people into geniuses, and meet the absurd standards of everyone who is unhappy, through very careful selection of opportunities, or a mixture of a significant, relaxed and composed, paradigmatic (emporium), and activities-at-every-level mentality, with an ability to easily differentiate landscapes based on major preferences. Also related to economics, which is all about the gift of metaphysics and its potential. Some people may be qualified to receive the gift. In fact, all may be qualified to receive some gift. So, gifts should have potential.

- To learn from the microcosm model in which difficult laws permit one or more micro-universes to be perfect in unison.
- To make Matter, Minds, and Meaning paradigmatic by law. In other words, to have a desirable metaphysics, where everyone has the ability to engage in an immaterial alchemy resulting in immediate benefits, e.g. qualities of meaning, in the environment.
- To provide very appropriate tools such as magical interactions, mutual ideas, imagination, and magical creator / design tools or mental aesthetics to permit realization of meaning.

3.

- To sustain the complexity of ideas in

the world as a standard of meaning. To use each meaningful idea to standardize all the others. To make peculiarities of meaning into facets of nature. To make the experience of meaning subtle, tangible, or fulfilling.
- To provide not just a metaphysical continuum, but a continuum of various ideas, as well as objects that demonstrate their meaning with fascination.
- To have meaningful qualities, meaningful properties, meaningful modes, and meaningful systems: creating a meaningful world with creatures capable of engaging in a meaningful existence.
- To 'make meaning wealthy'. To have significant intelligence and high quality in all things. To permit essences, substances, composites, and constructs with a view towards the metaphysical and ideal.

4.

- To have sufficient explanations for meaning, e.g. it's justice, preferability, uniqueness, naturalness, and so on.
- To attract people to the world who imperatively desire it's benefits, but are not incompetent. This may involve creating worlds with degrees of maturity corresponding to ease of use (less mature is easier and less emotional). Also, making the world function for those that benefit by it by taking sug-

gestions and interpreting them carefully. For example, was something universally said? Is there some exception? Does it apply to emotions more than truths? Does it result in cleverness? Can the cleverness be used meaningfully?
- To remember to guarantee some benefits, so the world doesn't go to waste. For example, fulfilling appearances, introspective meaning, mild euphoria, or a visionary quality. It may help to see the visionary as a cheap alternative to visual fulfillment, happy emotions, and scintillating thoughts.
- To permit meaningful realization, literal transformation, mood shaping, eternalism, and ideal paradigms.

Additional rules:

- No additional option without everything being made of souls.

Basically: recognizing relatively eternal work on information has occurred, updating it, having high standards, and using good ideas.

PROGRAMMABLE HEURISTICS AND BIG IDEAS

OTHER SYSTEMS /

PHYSICAL PARADIGMS:

Systems = Failed ambitions.

Expensiveness = Derivation of infinity.

Relativity = No infinite toughness.

Principle = No sacrifice.

Perfection = Efficient complexity.

Complexity = Slave to consciousness.

Epiphany = Motive for intellectual greatness.

Greatness = Justice.

Intellectual = Having ideas.

Pleasure = Derivation of theory.

Problems = Excessive ambition.

Ideas = Information about epiphanies.

OTHER SYSTEMS /

THE CALCULUS OF FREE WILL:

Preferences show an ability to meet wish with fulfillment. Rational behavior based on preferences indicates that willful behavior is not a matter of perception. Free will exists in the sense of the preferences, and rational behavior shows that preferences correspond to differences of volition.

If we cannot get what we prefer, or if we do not prefer anything, then determinism may be in play.

Free will is measured by the preferences, and cannot exist without them. Free will is expressed as timeless fulfillment. Free will is 4-dimensional.

As a thought experiment consider one has been denied an opportunity. As far as the preferences, in the material world it is the substance of the denial not the denial itself that creates a problem for the will. A material denial could be solved by other materials, or just saying 'this is magically okay' to yourself. Something else might be equivalent sometimes. The problem with the will is the discreteness of options, the inability to magically find an equivalent alternative, and the absolute denial of opportunities.

PROGRAMMABLE HEURISTICS AND BIG IDEAS

Note: There are four freedoms I know of:

The Freedom for the good (options).

The Freedom for everything (freedom).

The Freedom for judgment (deliberative activity).

The Freedom for imagination (possibility of thought).

In other words: preferences, context for preferences, ability to edit preferences, and research.

And the king of all preferences is another one: the ability to determine a time or fix a result.

Nathan Coppedge

PROGRAMMABLE HEURISTICS AND BIG IDEAS

SECRET LAWS /

LAW OF RELATIVELY SAFE CONSCIOUSNESS:

Individual ethics is something like this:

<u>What is general ethics?</u>

Acting for the greater good is when you are able to act ostensibly for the benefit of the whole: the wider context mostly not just yourself, under the assumption that others benefit in the same way you would in their position.

A critique of this is the radical, sometimes stupid-sounding metaphysical view that subjectivity and other people are intrinsically different things, thus greatly privileging the existence of an internal perspective as though an inhuman God—neither precisely internal, nor precisely external, had created it by design, everything being exactly as it appears, with internals and externals being separate and unrelated things.

SECRET LAWS /
THE MIND IS A WORMHOLE /
LAWS OF THE UNIVERSE:

[Coherent Systems 2.A.2.B.3.]

1

Given:

The Indefinite Condition, and

Arbitrary Structural

Then lim (phi) = infinite chain.

(…)

Infinite chain

+

Information whirlpools that are always circles (including all energy vectors)

=

Approximate material disintegral.

This is like saying time is a gigantic worm the size of the universe.

More on the first section here: Pursuing the Disintegral Part 2

PROGRAMMABLE HEURISTICS AND BIG IDEAS

2

We learned from the previous that the universe is a timeworm.

I have seen it myself.

Evidence that God can expand the universe is that it exists.

Evidence that God can compact the universe is wormholes.

There must be a third thing which is supreme but which can be found in intermediate particulate matter: consciousness.

However, coherently consciousness must be a unification of the expansion and contraction of the universe, since it is still part of the physics of timeworms.

Therefore, consciousness is not only the consciousness of being—the universe, but also a consciousness of wormholes.

From a traditional perspective then, consciousness is the consciousness of wormholes, and wormholes are the nature of consciousness.

This is a true unification of consciousness with the macroscopic properties of an incoherent universe involving timeworms.

This may be called the metaphysical trinity:

universe, wormholes, and consciousness, as a function of whatever the universe is, and modified by whatever conditions apply to what the universe is.

A simpler form is simply to say that 'the universe is its own animal, or it is an animal as animal is to something else'.

ADDITIONAL DETAILS:

> Maybe there is a law of uniqueness with organisms and universes.
>
> For example, if consciousness is more involved in the cosmos than thought, then consciousness might be a trinity with time / universe and wormholes.
>
> The equation of the disintegral predicts the universe is the snake of time, a very large snake, and consciousness and wormholes are the two other major elements. *The mind is a wormhole from the standpoint of the universe and other combinations.*

—Message to Tom Watkins on Quora, 2018/08/01

PROGRAMMABLE HEURISTICS AND BIG IDEAS

SECRET LAWS / MEANING /
HOW DO WE KNOW?

July 27, 2017.

> ***What some mean by meaning is all meaning. —Nathan Coppedge***
>
> —Souls of Analytic Philosophy

Why isn't everything we know wrong?

Because there might be something wrong with not knowing. For example, depression. Knowledge can be favored for practical reasons. Does this mean scream at your kid or start a revolution? Probably not!

Of what does knowledge consist?

Knowledge clearly has a component of meaning, since what we mean by knowledge is not meaningless knowledge. After all, if knowledge were meaningless we would have the same problem we had without knowledge.

We should at least know that knowledge involves practical meaning. For example, not screaming at your child or starting a revolution, as these things cause problems.

Nathan Coppedge

What is the limit of knowledge?

Knowledge, it must be said, is a concern for good things insofar as it is a practice of meaning to be concerned with that which has merit in itself.

Likewise, knowledge must in a similar sense not be concerned with bad things, insofar as those things must be meaningless to the good, or impractical to sustain.

Although knowledge is concerned with the good, that is, meaning and its practice, there is nothing which makes meaning inherently expensive or that should make us assume that when it is meaningful it must be impractical.

Thus, if madness is good it may be practical, if it truly is good, and one way for it to be good is if it is meaningful, and another is if it is practical.

But nor should we assume that all is madness, or all is practicality, for there may be many kinds of meaning which are good, and which are not yet practical or mad.

However, the principle of madness suggests that if it were truly good, it would also have meaning, and so the ultimate determination of meaning is on the basis of whether it is good.

And, as I said, what I mean by good is if it is good in every way. And what is meant by good in every way is if it is meaningful to the point of being practical.

And, in this way, the practical consists of an elevation of the good, that is, meaning.

What is meaning?

Meaning is that thing and anything which has a meaningful quality, and what is meant by this is something which signifies, not like the insane dolt that lifts a decapitated head, but rather more like the man who finds life to be full of a rich literature.

In fact, we can trace the path of higher meaning beyond mere words, and into the subjects like philosophy and art that are most pleasurable to those who do not risk meaninglessness, such as those who do not declare life to be meaningless.

(And these are the people who will not be depressed, and so they are relatively deathless).

Meaning then, is at least a lack of meaninglessness. And more than that, it is the elevation of practicality into those arts and literatures (or further concepts, modes, existences, etc) which suit themselves to the person who does not find life meaningless.

As such, the concepts are good, and so they have meaning, and so they are an elevation of practicality.

And where what is good must have meaning, so too what is meaningful must be the only elevation of practicality.

What is the ultimate?

The ultimate is any exceptional thing, which has its own meaning, good or bad, and as such can be judged ultimately practical or impractical.

It may be good and meaningful to have a home, just as it is good and meaningful to read a favorite book. Excitement is not the only reason to consider something as ultimate. Many ultimate things are in fact the boring things we ought to take for granted (because they have practical meaning).

Now, what is truly or extremely ultimate is just something we choose to consider highly exceptional, and this choice should maximize practical meaning.

There are techniques for doing this. For example, finding significance. Or strategizing for happy events. Many of these things, since they aim at the extreme, involve a kind of madness (exaggeration) or higher meaning (intelligence) bringing about the strategy or significance.

So, for example, there might be a practical and also an exaggerated view of an object, which puts it somewhere in the space of finding significance. In this manner, one may be connected to the idea of the sometimes extremely useful practical meaning of each object.

On the exaggeration of significance

It may be important to remember the importance of exaggeration in finding the meaning of objects. If something could be bad when it is exaggerated, it clearly does not have practical meaning, and so, it must be unrelated to the good, and so, it must be somewhat unethical to consider it. And when it is unethical to consider it, it is unlikely to involve itself in any kind of good life. And so, there is a sense of ethical virtues related to the exaggeration of practical meaning.

The ethical person is one who can exaggerate his every sensation and still find meaning. This means that everything is subject to critique, and it also means that if he is to be virtuous there will be no forgiveness, nor will there be any punishment other than meaninglessness. And so, we see it that the difference between the good life and the bad life is that the good life has significance, and the bad life is merely superficial, because the bad life is full of the meaningless things. In this way, we cannot take the bad life seriously, because there is no way that it is doing anything other than sharing in the misery of other bad lives. For if a victim of injustice were living a meaningful life, then there is nothing wrong with the thing that happens to him, and so the good life follows simply from the capacity for meaning, and the acceptance of universal exaggeration.

When one is honest, or even willing to exaggerate how bad things are, then this awareness prevents bad things from happening. It is similar to the awareness about bad things opposing significance.

When there is a desire to exaggerate the good things, the result will be perception of practicality and overwhelming meaning, which will together create practical meaning.

Thus, exaggeration is part of the good life.

Relevance for Meaning:

And when a thing has relevance it may be because it has a particular meaning for us.

Do… many 'odd conditions' expressly say that what we say has meaning does not have meaning? Obviously not, because these 'odd conditions' are part of the field of things that could be considered meaningful.

If it is something meaningless, it is not what is meaningful, and only the meaninglessness of the meaningful thing could contradict what is meaningful, so these 'odd conditions' never contradict what we find to be actually meaningful.

Therefore the relevance of meaning is a principle which goes higher than 'odd conditions' and confusing rationality, at least as far as the things relevant to meaning.

PROGRAMMABLE HEURISTICS AND BIG IDEAS

Furthermore, we have not said that what is relevant to meaning is absolute, but if we are justified in doing so, and it has a relevant meaning, there is nothing contramanding that.

SECRET LAWS /

EVERYTHING A LAW OF ITSELF:

> This seems somewhat amazing.
>
> A physical form of inference.
>
> A philosophical intuition.
>
> It's like mastering the atoms in front of your face.
>
> Things pile up in heaps and then you analyze them.
>
> Soul in italics.
>
> Then you can think about colored heaps, and it gets to be a lot of fun!
>
> —Rumors about natural deductions in the <u>Philosophy of Richard Volkman</u>

...

MAIN LINKS

Unified Theory types of deductions:

(1) <u>The Essential Proof</u>

PROGRAMMABLE HEURISTICS AND BIG IDEAS

(2) <u>Universal Proof of Natural Deduction</u>

(3) <u>Proof that Everything is Absolute by Natural Deduction</u>

(4) <u>The Universal Deduction of Nature</u>

(5) <u>The Laws of Nature</u>

[Coherent Systems 2.A.1.B.3.]

UNIVERSAL PROOF OF NATURAL DEDUCTION

[.: means 'therefore', U stands for universal, := means 'necessary condition that'.]

Some U

Substantial U

Law of U

.:

Law of some substance.

Some substance = Substantial U

.:

Some U := U

U : = Some

(U : = Some) := Some U

∴

U : = Some := Some U := U

Some := U

∴

Where Law is Some

Law = U

∴

If Laws must be Universal to be Laws,

And Some = Substance,

Substance is Law where there is Law and Substance and Some U

∴

Law == Universal Substance

PROGRAMMABLE HEURISTICS AND BIG IDEAS

SECRET LAWS /

THE UNIVERSAL EQUALITY:

Phenomenological insights into science, such as that

1. We are surrounded by virtual particles, real particles are larger,
2. The solar system is like an atom,
3. The universe is a worm made of consciousness, time, and wormholes,
4. The Higgs is significant like God,
5. The brain is a wormhole,
6. Electrons are the brain, a gold-like substance,
7. This all relates with Maxwell, indicating a kind of universal equality.

The first I know to realize this is **Katy Ruben**. *Something similar to* A Formula for Advanced Topics *may have been a prereq.*

See also: Rule of Equinimity

RULE OF EQUINIMITY:

Anything good is real.

A rule which shows equinimity, because it says that what trouble there is serves a purpose which lies in the future.

Nathan Coppedge

SECRET LAWS /

UNIVERSAL GREATNESS THEORY

PROGRAMMABLE HEURISTICS AND BIG IDEAS

SECRET LAWS /

OTHERWISE TRANSCEND

[Imagine a Transcendent Work of Art]

When in doubt, transcend.

When NOT in doubt, transcend.

Everything is basically transcendance.

Transcendance does everything, if there is anything left over, it requires transcendance.

Nathan Coppedge

SECRET LAWS /

DIMENTIA, EXCEPTIONAL HAZARD

PROGRAMMABLE HEURISTICS AND BIG IDEAS

SECRET LAWS /

IDEAL NUMBERS / A NUMERIC IDENTITY THEORY OF EVERYTHING:

Mistakenly called A Numerological Theory of Everything and Also Called Phenomenumerology;

Phenomenumerology;

ABOVE: The pertinent symbol of phenomenology.

The philosophy here is that complexity is content, and it requires more than one, but less than infinity: a god-game.

- Empty 1.
- Combined 2.
- Transit 3.
- Link 4.
- Metaphor 5.
- Game 6.
- Category 7.
- Collection 8.
- Book 9.
- Direction 10.
- Figure 11.
- Limit 12.
- Selfish 13.
- Network 36.

…

PROGRAMMABLE HEURISTICS AND BIG IDEAS

SECRET LAWS /
DOMINANT METAPHORS /
SECRETS OF THE WORLD:

1. Warren G Buffett's secret is all men are babies.
2. The powers will tell me things like people are reborn every 14 years after birth no matter what.
3. English is a joke about the Logos. Logos is actually inherently visually complex.
4. The Language of the Magi is like a joke about the Special Olympics. Rumor is you're not supposed to write the language of the Magi, only memorize it.

SECRET LAWS/

ORANGES MIGHT BE MADE OF GOLD/
THE GOLDEN RIND:

Formerly called "Food for Thought"

We can imagine peering slightly beyond the ordinary veil of existence, like seeing piles of gold beyond a picnic table.

What if philosophy = a broken system?

Then we would conclude that something much better lies outside of philosophy (is it gold, or instead *looking at the gold with no sense of ownership?* This could say things about the soul.)

Philosophy, as it were, is a particular variety of food (food, not gold, although it is very reminiscent of gold), a kind of food that happens when intellectual matters are 'broken', 'parsed', and 'distinguished'. "Maybe it does require a gold platter" says the philosopher…

Perhaps even the matter beyond philosophy is something which itself could be discredited, leaving something true, ebullient, and universal! (After all, the most precious metal is not gold!).

And yet, in my own particular way I find broken philosophy to be highly habituating, and so it seems there must be some connection be-

PROGRAMMABLE HEURISTICS AND BIG IDEAS

tween these two or more realms (the questions of what is *beyond* philosophy, and what *is* philosophy) if we are to find what philosophy really IS, or ISN'T.

With such ambiguity we might try to argue that there is an equally valid argument on any subject. The Equal Arguments by Nathan Coppedge on Official Nathan Coppedge Blog

Or even to make something concise and elegant such as mathematics or meta-philosophy which would perpetuate philosophy, whether it is broken or not, whether it even is philosophy or not!

That's my brain teaser for the day!

SECRET LAWS /

SCIENCE FROM EMPTY SETS:

[Coherent Systems 2.A.1.A.3.]

If I brought up the idea that proofs were possible under naive realism, most philosophers would say that defies the point, and it is against sanity.

However, even if that proves to be accurate, it is worth venturing a guess, because it might also prove useful.

1. There is proof under naive realism if it meets a naive-realist standard NS*, which is like an empty set with an empty conditional. The empty set with the conditional is ordinarily incapable of expressing anything, including zero.
2. But say that the conditional represents the condition under which the condition of naive realism is met. If this could be satisfied under naive realism, then naive realist proofs would be possible. And basically because we are dealing with naive realism, we can assume without analysis it means the same thing as an empty set with no implication.
3. But there is a slight problem with that view. Because, if naive realism can meet the standard of naive realism through a conditional indicating a na-

ive-realist standard, then there is a standard of naive realism, and proof is possible.
4. Now, how is knowledge impossible if an empty set can be proven by an empty conditional placed upon the set? What is wrong then with the empty conditional proving the naive realism of the empty conditional, since it refers to the same empty set?
5. Now, consider another case. The empty conditional is post-rational (what I call irrational) and empty, while the empty set is empty and assumed to be naive realist. Although this might be a broader case than using a naive realist conditional (post-rational includes both rational and irrational cases), and adds no analysis about truth-value, we know it is even easier to meet a post-rational standard of proof than a naive realist one. And given that the set and the conditional are empty, there is nothing against us doing so. In fact, there is no reason for naive realism not to support the use of post-rationality, particularly since it represents a broader case, which must therefore be more parsimonious by using fewer restrictions. So it turns out it is even *easier* to prove a naive realist standard by using parsimony. Now we must conclude that parsimony does not reduce the provability of naive realism.
6. However, if it is easier to prove naive

realism using parsimony, wouldn't we naturally extend this to all other systems which are generally considered less parsimonious? Indeed, if NS* proves naive realism under post-rationality, this indicates that everything standing on top of naive set theory is also proven.

Conclusion: *Everything is proven that has a standard. Relative to absoluteness is not relative.*

[Nathan Coppedge's Response to Beatrix Esmond on Justified True Belief:] What would Esmond say about the concept of multiple truths? Does multiple truth imply contradiction? And how can we refute truth in general without refuting multiple truthism? It seems to me that if we do not reject multiple truthism, there may indeed be some truth or collection of truths, which, if contradictory or not, produces the result that a given statement is true under some conditions, and there is really nothing saying those conditions cannot be met conditionally. Furthermore, there is really nothing preventing us from accepting multiple truthism completely, because: 1. It may express all truths, 2. Some of them are bound to be right, and 3. Even if some are contradictory, we might do better to accept some conditionally rather than rejecting all unconditionally (at least, in my opinion).

SECRET LAWS /
STANDARD THAT SCIENCE IS NOT SCIENCE / THE FALLACIOUSNESS OF SKEPTICAL DECONSTRUCTION:

(1) The common skeptical argument will begin by denying examples like absolute knowledge, and this cannot be true under relativism.

(2) so, the scientist cannot be a relativist.

(3) But one cannot be skeptical of everything.

(4) For we know the utter solipsist is also a relativist, or something very similar.

(5) So, the scientist is defending something. And, yet, the scientist claims to be a skeptic.

(7) But he is not a relativist, and so, in some views, he should not be a skeptic.

(8) So, what does it mean to be a skeptic who is not a relativist? There is something arbitrary about this claim.

(9) And, where the skeptic argues the universe itself is arbitrary, how can he argue knowledge is not arbitrary?

(10) *Remember that if the skeptic is a relativist, he cannot argue against absolute knowledge if such exists, as relativism can always be

relativized again.

(11) On one hand we have the scientist who says he is not a relativist, and on the other we have the scientist who says the universe is arbitrary. How is this not relativism?

(12) Now, we could conclude scientists do not believe the universe is random, or that scientists are relativists, or that scientists are naive realists and think things are just as they appear with no opportunity for translation, or that they place knowledge ahead of science.

(13) Of these, the claim that knowledge is ahead of science is the most fantastical for a skeptic to believe. But one of the others, that the universe is not random, might be a religious claim.

(14) So, for a scientist, the most appealing positions must be solipsism and naive realism (the realism must be naive because if not we must favor a priori knowledge over empiricism). And so, if the scientist knows anything, he is contradicting the initial position that science is not relative.

(15) Now the scientist will have to evade relativism to make a knowledge claim.

(16) *True* naive realism will not be an option if science is making a knowledge claim at all, and so the scientist will need to relativize by qualifying terms if he wants to be naive realist at all.

PROGRAMMABLE HEURISTICS AND BIG IDEAS

(17) Now the scientist turns to his third best choice, which is that the universe is not random. But this is not a particularly skeptical position. What makes the scientist feel that science is so important, if the universe is not random? This looks like a religious egoism.

(18) Now, science turns to the last position, and by now clearly science has lost. All that is left is to see philosophy and science on equal terms.

(19) But, if philosophy and science are on equal terms, clearly empiricism is not the only way.

(20) And so, science is not objective.

(21) And so, science is not deconstructive.

(22) And so, we might guess science is not reductive.

(23) And so, we might guess science is not a science.

...

Notes

Closest patent match I could find was (DAG): <u>WO2002046980A2 - A method of configuring a product using a directed acyclic graph - Google Patents</u>

(I don't advise buying rights to this unrelated patent unless you know what it actually does. It's probably just a generalized legal defense)

PROGRAMMABLE HEURISTICS PART 2

(The same restrictions hold: cite the author).

ABBREVIATIONS

R = Invaluable Proof and **R**efutation. Refuting by streamlining necessary conditions. For example, if a space station was founded on unethical principles, it will remain unethical, or it was not effectively unethically founded. Beginning with lemma and ending with doubt of that lemma, contingent on truth. <u>The Invaluable Proof and Refutation</u>

C = **C**olundrum. Problematic condition leads to a negative correlation of related condition. For example, bad philosophy produces pragmatism where pragmatism opposes philosophy. A means of predicting rules without assuming systematic properties. <u>High-Minded Colundrum</u>

Poly = **Poly**calculus. Physics... number theory... qualified or unqualified. A means of forming an intelligent calculus on any subject. <u>The Polycalculus Concept</u>

Str = **Str**ing-Theoretic Deduction. X →String Theory → Space Time. A means of deducing advanced physical theories. <u>String-Theoretic Deduction</u>

HEURISTICS:

Poly → Str (= *converting between philosophy of physics and physics*).

R → Lemma → Elaborate → Q (= *targeted questions from premises*).

R → Lemma → Poly → Str (= *Elaborate system from provable lemmas*).

R contra C → Elaborate Str. (= *elaborate world from non-contradiction or problematics*).

Psy contra R → Constant (= *prediction without a lemma means no change in the system*).

C contra R → Exceptional set.

See also the larger set, at: Nathan Coppedge's answer to In what ways has the "Linguistic turn" proven fruitful in Western philosophy?

PROGRAMMABLE HEURISTICS AND BIG IDEAS

BACKGROUNDS FOR PROGRAMMABLE HEURISTICS PART 2

THE INVALUABLE PROOF AND REFUTATION

[Coherent Systems B.4.A.3.]

Is it for common benefit? (For example, is it possible that Krishna's mind-island is causing a headache?)

If not, what is expressed? (For example, is it causing a headache?)

What can be done? (For example, should the headache disappear?)

As a result of this, what is it that's true? (That Krishna's mind-island causes a headache if it is borrowed from someone else's mind).

See also *Socratic Writings* (Coppedge 2017).

HIGH-MINDED COLUNDRUM

Colundrum: A high-minded solution to a conundrum. It means something like 'disassembling the doctor'.

GENERAL FORM:

Conundrum 1: The more we analyze, the more we criticize analysis.

Colundrum 1: Higher Principle: The less we analyze badly, the less we have problems with analysis (analysis may be bad without analysis).

Conundrum 2: Love is a vain sport.

Colundrum 2: A bad lover will not sport with vanity (the lover may not be so bad).

Conundrum 3: War is inevitable in political conflicts.

Colundrum 3: Peace is not inevitable in the absence of political conflict (peace may not be inevitable).

Conundrum 4: Philosophers ask too many questions.

Colundrum 4: What is not philosophical does not ask many (philosophical) questions.

POLYCALCULUS

[Coherent Systems D.1.C.1.]

Inspired by previous works, such as Clues About Coherent Calculus and System 2, I set about to design a concept of multiple-calculus or polycalculus.

An initial clue is that polycalculus relates to 'many worlds' of calculus. This could be

PROGRAMMABLE HEURISTICS AND BIG IDEAS

equated for example, with adding sigmas.

Another interpretation is that other logics could be used which are less about sums and more about modifications between perspectives. Although some of these would produce a perspective that is pronoucedly linear, others would produce differentiation or layers that may be either finite or infinite.

Thus, initially three categories exist: non-valued, quantitative, and infinite. In a wider sense, the system also accepts modifications of these. So, for example, quantitative could become arbitrary, ambiguous ... number theory. Non-valued could be evaluative, transitional, physical... etc. Infinite could have a qualified or unqualified standard.

The question becomes, how to make math out of all this? In one sense the outcome might be irregular and disorganized. In another sense it might be structured by the involved elements.

Physics... number theory... qualified or unqualified appears to be the linear, quantitative approach to polycalculus. This assumes some slant towards coherence.

However, we can add to this concepts which extend the method and permit greater control of the results:

Symbols... physics... number theory... unqualified... qualified...set theory... megaverses...

For better use, we might seperate these into

pairs:

1. Symbolic physics.
2. Unqualified number theory.
3. Qualified set theory.
4. Megaverses.

At this point it should be possible to grasp something of what is meant by polycalculus: A branching structure taking a wide variety of well-chosen operators, in which the expression of the result is open-ended, and may begin or end in multiple places and for multiple purposes, always incorporating some of the above elements. Thus, polycalculus can be simple or complex, parsimonious or ersatz, symbolic or naive, rigorously mathematical or even be a single logic operator. There is nothing about polycalculus that makes it inherently organized or rigorous, except that it's structure provides a general format for finding (with some effort) all forms of calculus.

POLYCALCULUS EXTENDED (E.G. PAPER ON POLYCALCULI)

FIRST INVESTIGATION INTO A MAJOR CLASS THEORY OF THE POLYCALCULII

ABSTRACT: This paper will investigate the general class theory of polycalculii, and theory originally developed and introduced on the Quora platform as a kind of philosophy of logic and mathematics. The ostensible goal at minima will be to demonstrate the exclusion of an array of class types which edicate the theorem of the polycalculus. The polycalculus is a general method for deriving equivalent calculus's other than the traditional intelligent calculus. The domain of this discussion is not only mathematics, but also logic, philosophy, and philosophy of science, as well as philosophy of the foundations of mathematics.

I. MISSION:

The goal of this brief project is to demonstrate a variety of coherent modes of inference alluding to the expandability of mathematics and the capacity to generalize mathematics over a relatively larger logical domain. The central problem will be to elucidate a central 'motive process' to calculation. A motive process shall be in this case simply a more fundamental process that still allows for the respective tools of each constituent calculus without eliminating functions of any calculus. However, the full proof that functions have not been eliminated is beyond the scope of this paper. In order to be parsimonious, there will be an operant insight implicated in each calculus that its fundamental operability is retained. Each calculus (or LOGIC) will be seen as representing a fundamental operability of a distinct type, although this distinct type will not always be elucidated except by mentioning the type of calculus.

II. STRUCTURE / DEFINITIONS

Polycalculus was previously disclosed to have the following general structure:

> *Physics* →
> *Number Theory* →
> *Qualified or Un-Qualified*

Thus, it is obvious that with the binding assumption that this is a correct operation, providing a permutation of appropriate elements in each part of this series would lead directly to a complete list of calculi. One other assumption assumes that each of the three parts contains all possible elements within its part, or that the mode of the part within the overall structure of calculus is understood. One should assume at this point that the formula is accurate in order to deduce the appropriate results.

III. REPRESENTATION:

Part 1.
Physical Calculus. This may be seen of consisting of the exclusive elements of experience, which may be taken to be Infinity, Relativity, Complexity, and Infinitesimal Calculus.

Part 2.
Number Theory. This may be seen as consisting of Set Theory, Cardinals subsumed within Rationals, Proportional Trans-Finites, and Extended Numbers.

Part 3.
Qualified or Unqualified

This is the distinction between Intelligent and Literal Calculus, implicating an element of calculus in both cases. Literal Calculus is for example the limit as it approaches zero, the anti-derivative, the normalized wave functions of information theory, and the change of a system (t) in chaos theory. Intelligent Calculus is for example, intuitions on advanced physics, medicine, and chemistry.

The result of the permutation is 4 X 4 X 2 = 32 Calculii.

Calculus 1: Infinite Set, Intelligent

Applications of categories, complexity, algorithms, paradigms.

Calculus 2: Infinite Set, Literal

Exhaustive processing, exclusion principles, evolutionary processes.

PROGRAMMABLE HEURISTICS AND BIG IDEAS

Calculus 3: Infinite Cardinals Subsumed Within Rationals, Intelligent

Mathematical intuition, I.Q., functional heuristics, adaptogenius.

Calculus 4: Infinite Cardinals Subsumed Within Rationals, Literal

Knowledge of limits, Laplacian, Gauss curves, chaos theory, probability.

Calculus 5: Infinite Proportional Trans-Finites, Intelligent

A Priori, Deterministic, Ethical, Aesthetic.

Calculus 6: Infinite Proportional Trans-Finites, Literal

Extension, Variation, Creativity, Repetition.

Calculus 7: Infinite Extended Numbers, Intelligent

Absolute, Complete, Representative, Theoretical.

Calculus 8: Infinite Extended Numbers, Literal

Direct, Political, Magic, Natural.

Calculus 9: Relative Set, Intelligent

Elemental, Deductive, Formal, Semantic.

Calculus 10: Relative Set, Literal

Artistic, Figurative, Strategic, Physical.

Calculus 11: Relative Cardinals Subsumed Within Rationals, Intelligent

Large Sums, Economics, Prophecy, Number Theory.

Calculus 12: Relative Cardinals Subsumed Within Rationals, Literal

Variables, Arithmetic, Spatial Reasoning, Conjunction.

Calculus 13: Relative Proportional Trans-Finites, Intelligent

Meta-Mathematics.

Calculus 14: Relative Proportional Trans-Finites, Literal

Vector Analysis.

Calculus 15: Relative Extended Numbers, Intelligent

Medicine.

Calculus 16: Relative Extended Numbers, Literal

Conjectural Analysis, Meditation, Un-Attachment, Parsimony.

Calculus 17: Complex Sets, Intelligent

Invaluable Proof & Refutation, Computational Inference, Noble Deduction, Cutting the Gordian Knot.

Calculus 18: Complex Sets, Literal

Problematics, Stating the Problem, Case Studies, Prior Art, Critical Methodology.

Calculus 19: Complex Cardinals Subsumed Within Rationals, Intelligent

Mazery, Dialectical Reasoning, Guesswork, Emergent / Emergency Thought.

Calculus 20: Complex Cardinals Subsumed Within Rationals, Literal

Exploration, Adventure, Procedural Reasoning, Empiri-

PROGRAMMABLE HEURISTICS AND BIG IDEAS

cism.

Calculus 21: Complex Proportional Trans-Finites, Intelligent

Recursive, Operational, Synchronizing, Retrofitting.

Calculus 22: Complex Proportional Trans-Finites, Literal

Organizational, Informational, Neutral, Symbolic.

Calculus 23: Complex Extended Numbers, Intelligent

Transcendental, Visionary, Impresario, Beautiful.

Calculus 24: Complex Extended Numbers, Literal

Exceptional, Significant, Inherent, Complex-Ordinary.

Calculus 25: Infinitesimal Sets, Intelligent

Egregious reasoning.

Calculus 26: Infinitesimal Sets, Literal

Formal Aesthetics, Minimalism, Justice.

Calculus 27: Infinitesimal Cardinals Subsumed Within Rationals, Intelligent

Data, Proto-Semantics, Notation, Lemmas.

Calculus 28: Infinitesimal Cardinals Subsumed Within Rationals, Literal

Formatting, Computer Language, Markup, Legal Style.

Calculus 29: Infinitesimal Proportional Trans-Finites, Intelligent

Essential, Potential, Demonstrative, Elective.

Calculus 30: Infinitesimal Proportional Trans-Finites, Literal

Mechanical, Psychological, Spiritual, Animation.

Calculus 31: Infinitesimal Extended Numbers, Intelligent

Broken, Ugly, Conceptual, Universal.

Calculus 32: Infinitesimal Extended Numbers, Literal

Situational, Synthesizing, Found Object, Emotion.

To summarize:

1. Algorithm, 2. Exclusion, 3. I.Q., 4. Limits, 5. A Priori, 6. Extension, 7. Complete, 8. Direct,

9. Formal, 10. Artistic, 11. Number Theory, 12. Variables, 13. Meta-Mathematics, 14. Vector Analysis, 15. Medicine, 16. Conjectural Analysis,

17. Gordian Knot, 18. Problematics, 19. Emergent Thought, 20. Adventure, 21. Operational, 22. Symbolic, 23. Beautiful, 24. Complex-Ordinary,

25. Egregious, 26. Justice, 27. Lemmas, 28. Formatting, 29. Essential, 30. Psychological, 31. Conceptual, 32. Synthesizing

IV. PROPERTIES:

If it is seen that one calculus transforms into the next due to their related properties, then when X calculus follows Y calculus we can say Y has a property of vector X or VX. If X calculus precedes Y calculus we can say X has a priority on Y, or Y has a property of PX. So, the 32 calculi can be expressed in this form as follows:

PROGRAMMABLE HEURISTICS AND BIG IDEAS

1
Synthesis P Algorithm V Exclusion. (The principle of an algorithm is to exclude synthesis).

2
Algorithm P Exclusion V I.Q. (The I.Q. of an exclusion is not an algorithm).

3
Exclusion P I.Q. V Limits. (The limit of an I.Q. is not an exclusion).

4
I.Q. P Limits V A Priori. (The A priori limit is not I.Q.).

5
Limits P A Priori V Extension. (The extension of the A priori is not the limit).

6
A Priori P Extension V Complete. (The completeness of extension is not A priori).

7
Extension P Complete V Direct. (The direct reason of completeness is un-extended).

8
Complete P Direct V Formal. (The formalism of directness is not complete).

9
Direct P Formal V Artistic. (The artistic formalism is not direct).

10
Formal P Artistic V Number Theory. (The number theory of art forms is not formal).

11
Artistic P Number Theory V Variables. (The variables of number theory are not artistic).

12
Number Theory P Variables V Meta-Mathematics. (The meta-mathematics of variables is not number theory).

13
Variables P Meta-Mathematics V Vector Analysis. (The vector analysis of meta-mathematics is not variables).

14
Meta-Mathematics P Vector Analysis V Medicine. (The medicine of vector analysis is not meta-mathematics)

15
Vector Analysis P Medicine V Conjectural Analysis. (The conjectural analysis of medicine is not vector analysis).

16
Medicine P Conjectural Analysis V Gordian Knot. (Cutting the Gordian knot of conjectural analysis is not medicine)

17
Conjectural Analysis P Gordian Knot V Problematics. (Problematics of Gordian knots is not conjectural analysis)

18
Gordian Knot P Problematics V Emergent Thought. (Emergent thought of problematics does not cut a Gordian knot)

19
Problematics P Emergent Thought V Adventure. (Adventure of emergent thought does not have problematics)

PROGRAMMABLE HEURISTICS AND BIG IDEAS

20
Emergent Thought P Adventure V Operational. (Operations of adventures are not emergent).

21
Adventure P Operational V Symbolic. (Symbolic operations are not an adventure).

22
Operational P Symbolic V Beautiful. (Beautiful symbols are not operations).

23
Symbolic P Beautiful V Complex-Ordinary. (Complex-ordinary beauty is not symbolic).

24
Beautiful P Complex-Ordinary V Egregious. (Egregious complex-ordinariness is not beautiful).

25
Complex-Ordinary P Egregious V Justice. (Justice that is egregious is not complex-ordinary).

26
Egregious P Justice V Lemmas. (Lemmas that have justice are not egregious).

27
Justice P Lemmas V Formatting. (Formatting of lemmas is not justice).

28
Lemmas P Formatting V Essential. (Essential formatting is not lemmas).

29
Formatting P Essential V Psychological. (Psychological essentials are not formatting).

30
Essential P Psychological V Conceptual. (Conceptual psychology is not essential).

31
Psychological P Conceptual V Synthesizing. (Synthesizing concepts is not psychological).

32
Conceptual P Synthesizing V Algorithm. (Algorithms of synthesis are not concepts).

To summarize again, full knowledge of the 32 poly-calculii might amount to:

The principle of an algorithm is to exclude synthesis. The I.Q. of an exclusion is not an algorithm. The limit of an I.Q. is not an exclusion. The A priori limit is not I.Q. The extension of the A priori is not the limit. The completeness of extension is not A priori. The direct reason of completeness is un-extended. The formalism of directness is not complete. The artistic formalism is not direct. The number theory of art forms is not formal. The variables of number theory are not artistic. The meta-mathematics of variables is not number theory. The vector analysis of meta-mathematics is not variables. The medicine of vector analysis is not meta-mathematics. The conjectural analysis of medicine is not vector analysis. Cutting the Gordian knot of conjectural analysis is not medicine. Problematics of Gordian knots is not conjectural analysis. Emergent thought of problematics does not cut a Gordian knot. Adventure of emergent thought does not have problematics. Operations of adventures are not emergent. Symbolic operations are not an adventure. Beautiful symbols are not operations. Complex-ordinary beauty is not symbolic. Egregious complex-ordinariness is not beautiful. Justice that is egregious is not complex-ordinary. Lemmas that have justice are not egregious. Formatting of lemmas is not justice. Essential formatting is not lemmas. Psychological essentials are not formatting. Conceptual psychology is not essential. Synthesizing concepts is not psychological. Algorithms of synthesis are not concepts.

PROGRAMMABLE HEURISTICS AND BIG IDEAS

V. PREDICTIVE POWER

The model only holds under the condition that the relevant observations can be explained within limits, or the extension of limits into other calculi. The model does not account for anything except calculus, and even then only the calculus defined as knowledge of mathematical limits and their extensions. Nor does the model yet provide for the exact mathematics of these derivatives, if they are indeed mathematical. Likely a much wider range of mathematical operations would be necessary to express the full set of calculi.

VI. EXAMPLES

The application of the knowledge is somewhat self-evident, although in its present state it might take the form of a heuristic.

VII. FALSIFIABILITY

If the permutation is exclusive, and with sufficient insight, then the entire set is indeed a type of complete description of calculus. I invite readers to investigate any shortcomings of the analysis. However, doing so may require creativity.

VIII. BACKGROUND

The original, very short formula for Polycalculus was created on March 10, 2017 by Nathan Coppedge after a series of inspirations. His writing on Polycalculus appeared on Quora on that day.

[CONTINUED BACKGROUND FOR PROGRAMMABLE HEURISTICS PART 2]

STRING-THEORETIC DEDUCTION

[A development of Space-Time-Gravity Theory and String Theory, as well as Polycalculus]

A.

X → String Theory → Space Time.

For example,

1. Quantum location →Equal energy particle theory → P Branes.
2. String theory → Quantum space-time →Relativity.
3. Time → Affected time strings →Quantum location.
4. Quantum tunneling → Tensor flow →Hive brane.
5. Quantum momentum→ Reverse particle theory → Spin electrodynamics.
6. Negative entropy → Quantum formation →Hypostatic gravity.

B.

What's real integrates with a functional integrator.

1. Integration disintegrates.
2. Disintegration integrates.
3. Relativity.
4. Superstring theory.

PROGRAMMABLE HEURISTICS AND BIG IDEAS

ADVANCED PROGRAMMA-BLE HEURISTICS

This group assumes hardware can be as valuable as software.

Advanced heuristics generally concern the superficial and the metaphysical aesthetic.

Like:

Meaning + Efficiency = Good (in terms of energy).

Sacrificing what isn't everything for something → that's how we get inefficiency. (Sacrifice = Inefficiency).

Exponential efficiency.

Intellectual materiel.

Hardware-software.

Value integration.

Careful formatting.

Non-arbitrary interaction with arbitrary elements.

Dimensional modelling.

Integrated programs.

Formula writing.

Snap formulas.

Formula integration.

Paradigmatic function (s).

Ideal situations modelling and derivatives.

Elevated problem cases.

Aesthetic logic.

Metaphysical paradigms.

Philosophic shortcuts.

Minimal cohorts (back doors and other mysteries).

Exeter methods (causal and logical logistics over minimal cohorts).

Formal tools / Recursive archetypes.

Star patterns / Quantum processing.

Isometric organizers / substitutional logix / parallel quantum processing.

Functional models / perpetual motion computers / perfect evolutionary neural networks.

1-d advanced processing.

2-d advanced processing.

Fractional advanced processing.

4-d pure computation.

PROGRAMMABLE HEURISTICS AND BIG IDEAS

5-d pure computation.

8-d pure computation.

16-d semantic computer.

Qualified creative computing.

Perfect selective logix.

X-dimensional computing.

Objective computing 1-degree, qualified computing, omni-computing.

ENGLISH HEURISTICS

Answers using these formulas seem always to be right, for obscure reasons.

1 .Terrapinisms.—

While they were floundering,

I was pondering...

No more wandering through the dark tunnels of grim determination—

For No!

It is time to grow in a thousand-folded folds!

For which we need an infinite fuel!

Var:

While they were foundering,

I was floating...

No more wandering through thoughts that lack the jest of discrimination—

For know:

It is a metaphor with a million folds!

For this we need a finite fool!

2 . Rhyme of Variation—

A walking headache against dementia,

PROGRAMMABLE HEURISTICS AND BIG IDEAS

His invention was perpetual!

Var:

A firm headache with resolution,

His answer was the solution!

3 . Using What-Is-Not Method

Take a four part set of opposites opposed along the diagonal. Now iterate each category three more times, forming the corresponding parts of three further sets. Note that this doesn't always work, and may take some skill.

One example is (1. Mystery, 2. Origin, 3. Evidence, 4. Proof, 5. Resolution, 6. Exception, 7. Opportunity, 8. Determination, 9. Cause, 10. Connection, 11. Effect, 12. Disconnection, 13. Creation, 14. Construction, 15. Judgement, 16. Perfection).

Another example used straight permutation: (1. Ambition is fortune, 2. Evidence is error, 3. Failure is method, 4. Poison is principle, and the remainder is just a cyclic permutation moving the first to the fourth and moving the others up for each subset.)

Use the resulting 16 equal, well-ordered types (1.1, 1.2, 1.3, 1.4, 2.1, 2.2, 2.3, 2.4, 3.1, 3.2, 3.3, 3.4, 4.1, 4.2, 4.3, 4.4)

Can be re-arranged into a second, smaller, more efficient set by comparing opposites using what is not to join them (1.1 to 3.3, 1.2 to 3.4, 1.3 to 3.1, 1.4 to 3.2, 2.1 to 4.3, 2.2 to 4.4, 2.3 to 4.1, 2.4 to 4.2) forming two halves marked 1.1 to 2.4.

The resulting set can be further condensed using what-is-not (1.1 to 2.3, 1.2 to 2.4, 1.3 to 2.1, 1.4 to 2.2). The resulting set is simply marked 1 to 4.

It can be further condensed, using what-is-not (1 to 3, 2 to 4). The result is a set with two complicated statement-categories.

Now, again, 1 with 2, using what-is-not between 1 and 2. The result is a coherent, refined aphorism on the ultimate truth of the original categories.

The first example results in:

What is un-mysterious and opportune is that the effect of creation is that there is evidence of resolution in the causeless judgement: the original determination of connected constructions is that what is not proof of exception is connected perfection.

The second example results in:

When it is fortunate to be a failure in ambition, and principle fails, and method fails ambitiously, and error is not opposed, then error is in manifest evidence, and fortune does not poison truth; principle is in evidence, and method may or may not give evidence.

4. Extemporaneous Methods

Examples to learn from:

If the end of an infinite parser process is creation, we might interpret that: *Endless iterations were endless conclusions, absolute for-*

PROGRAMMABLE HEURISTICS AND BIG IDEAS

mulations.

Other examples,

Where greed or folly succeeds, something is evident or we cannot be ashamed. OR, *If we concede to greed or folly, the evidence is that there was some greedy fool.*

Where the economy of a labyrinth is that it is a labyrinth, the value of a labyrinth is only that it is a luxury.

(Casual Note: The last two sections are a condensement of parts of my book How To Write Aphorisms, published several years ago. The first quotation—about perpetual motion—originally dates to my website at Nathan Coppedge's Perpetual Motion from around 2008).

INFINITE-DIMENSIONAL HEURISTICS

- You don't know infinity, then you don't know infinity.
- You don't know infinity, then you don't know the calculus.
- You don't know infinity, then you don't make a prediction in higher dimensions.
- Complex truths are only infinitely perfect in infinite dimensions, or they involve more than one dimension simultaneously.
- If something infinite is genius it is a bit of a genius.

This is based in part on: <u>Observations on the Infinite Goldfish</u>

And has some additional correlated writings here: <u>Support for Holographic Consciousness</u>

PROGRAMMABLE HEURISTICS AND BIG IDEAS

INFINITE-DIMENSIONAL HEURISTICS PART 2

1. Everything exists between God and a trapdoor spider: impossible depth, and perfect height.
2. In between impossibility and perfection are many infinities.
3. The end is simple: now, then, there, why, etc. The details are relatively complex: matter, senses, intelligence, dimensions, etc.
4. What is, is. What is not, is not. Everything is. Everything that is not, is not. What is eternal returns. What is certain varies.
5. What is most clever depends on the thing, and what is eternally clever.
6. What is noticed is the least significant thing.
7. What is not noticed at all is beneath being noticed.

HEURISTIC SCHMOOZLE #1: GENIUS / WHAT IS THE PSYCHOLOGY OF GREAT MINDS?

I'll try to tackle this one.

1. Metaphor.
2. Philosophy.
3. Calculus.
4. Physics.
5. Innovation.

Philosophy of Physics type approach:

Nathan Coppedge's answer to Is it possible that at some point in time and space every form of intelligence converges towards the same conclusion?

Philosophy type approach: answer a question that has not been answered:

The Dialectic of Philosophy (Philosophy's Most Important Questions)

Basically, genius seems to be divided into Artists, Philosophers, Mathematicians, Physicists, and Inventors.

(Basically: Artists are trying to use metaphors. Philosophers are trying to figure out what philosophy is. Mathematicians are using Calculus. Physicists are doing philosophy of calculus. Inventors are interested in innovation).

PROGRAMMABLE HEURISTICS AND BIG IDEAS

There are also a few understated types including Writers, Activiists, Politicians, Psychologists, and Musicians.

(Basically: Writers are trying to find a better metaphor. Activists are trying to use philosophy. Politicians are trying to find something superior to calculus. Psychologists are trying to find something superior to physics. Musicians are trying to innovate the concept of innovation).

There are also a few even more understated types including Biologists, Chemists, Comedians, Doctors, and Priests.

(Basically: Biologists are tired of metaphors. Chemists are tired of philosophy. Comedians are tired of calculus. Doctors are tired of physics. Priests are tired of innovation).

By the way, my professor T.W. Bynum said to me once "I thought you should know all biologists were once musicians." I said that sounds a little too gay and I said thank you anyway it sounds interesting, and he said he knew that but he's not gay and said it came to him in a dream, but he didn't know what it meant.

That phrase (not my professor, but the language he used) that "All biologists were once musicians" was one of the things that inspired me to have intuitions about quantum mechanics, relativity, string theory, medicine, Platonism, and some other areas, and ultimately to admit that genius seems to consist of metaphor, philosophy, calculus, physics, and innovation.

Nathan Coppedge

AGING, HEALTH, AND LONGEVITY HEURISTICS

1. Age matters because time matters.
2. Don't be unwise (don't be swayed by common opinion, don't risk your life, and don't commit suicide).
3. Do what's good for you (like eating your fruits and vegetables, and exercising).
4. Keep an active mind.
5. Sometimes applying a basic heuristic can make a lot of difference.

For other secrets, see:

The first major clues I had were:

1. Adaptivity.
2. Age before youth.
3. Medicine.
4. Athleticism.

I also recommend:

- Avoiding artificial ingredients like food dye and fake sugar.
- Avoiding white bread.
- Avoiding diet sodas.
- Not building up too much muscle, as this can become fat when you stop exercising (most older guys have this problem, because they did weight-lifting).
- Cut down on fat consumption by the age of 25 to 35.
- Cut down on sugar consumption by the

PROGRAMMABLE HEURISTICS AND BIG IDEAS

age of 35 to 45.
- Avoid strenuous work.
- Do some of the things you want to do (browsing books, hobbies), as this will help you stay mentally active.
- Remember to save some small unpredictabilities for later. If you can be unpredictable when you're 60 to 120, this gives you a mental edge.
- Remember to be a little bit strong if you reach 100. Sometimes the only problem at that point is physical strength.
- At some point you will need to make big discoveries with herbal medicine. Probably the earlier the better. Jiaogulan, ginseng, and turmeric might help, along with uncooked spinach for general organ health, asparagus for bone marrow, broccoli against cancer, and fish (the edible part of the skin especially) for brains.
- If you want to live to 101 to 110 or beyond, you will probably want to know some magic. The kind of magic that old people have when they are still highly active.

I also recommend my book, THE DIMENSIONAL IMMORTALITY TOOLKIT which contains some advice from those who have searched for immortality.

As an added bonus, I'll throw in Nathan Coppedge's secret techniques:

Nathan's Immortality Techniques, Set #1:

(1) Make little bets with yourself about what something means. It could be whether you look at a car passing, or whether you choose

to smile or eat a snack at a given moment. If it turns out one way, you decide that it has a given result. So you use it to hone your willpower. Not only that, but you use it to gain motivation about what you can really achieve. You can tell yourself, 'my smile means immortality' or 'when I eat this snack, it's for a good reason, so I'm going to make some achievement'. Instead of being completely flexible, apply principle now and then, and use it as guidance for the long-term.

(2) See your own past as something that was infinitely old, and your future as something moderately young. This will allow you to justify endless good health. Life is not a sacrifice or something idiotic, but rather an endless struggle to regain lost wisdom, by becoming young. Youth will provide a platform for the infinite quest that is required.

(3) By your own definition, try to be lucid about your entire life whenever possible. This involves seeing through the mists of time that enshroud your life, and making some predictions about what will happen to you, and what your life means. This will make you relatively wise compared to people who do not do this.

(4) Do small things to improve your health incrementally as time goes on. By this logic, if you are infinitely old, you will definitely be infinitely healthy! Use this logic to convince yourself that you will never die, or at least, when your health leaves, it was a robber who took it, not someone giving charity.

Nathan's Immortality Techniques, Set #2:

PROGRAMMABLE HEURISTICS AND BIG IDEAS

(1) Don't Make Evil Bargains. Obey your best principle. Your soul or sense of self-motivation is the deepest underlying principle related to you and your life.

(2) Practice Virtuous Health. Health is the best kind of karma, because you get the reward immediately. I am a deep believer that when you are healthy, others are healthy too. Don't believe appearances about others' sickness or sense of well-being. Trust improvement in your own life, and only when it improves!

(3) Have Principle. Creating laws to live by --- that you really believe in --- can make a difference for your perception of time and the prospect of immortality.

(4) Mix Ingredients in Your Mind. Practicing a little mental chemistry can make a huge difference in the long-term. Remember, its not about drugs, but your own sense of self. Even if you don't notice results from your thought process immediately, remember that where you have a principle, you can improve eventually! Where you have headaches, you may later have happiness! Where you have blunt emotions, you may later become vibrant! Miracles are possible with a good brain!

Nathan's Immortality Techniques, Set #3:

(1) Permutations of interpretation. Infinite dimensions of time, forgettable, exposable.

(2) Periods of activity and inactivity.

(3) Ironic about the negative. An attitude of

delayed indifference.

(4) Ask originality of objects, leave it ambiguous if they do not reply.

Nathan's Immortality Techniques Set #4:

(1) Ritual.

(2) Distraction.

(3) Patience.

(4) Permanentize.

[End of Aging, Health, and Longevity Heuristics]

PROGRAMMABLE HEURISTICS AND BIG IDEAS

HEURISTICS FOR HAPPINESS

1. Meaningfulness: Noticing meaning.
2. Sadness → meaning → happiness.
3. Avoiding Stress While Attending to One's Needs: Caring for yourself combined with avoidance of stress.
4. Insanity: A little insanity, humor, relaxation, or words in your own defense.
5. Moderation without Suppressed Feelings: Moderate stimulation plus avoiding suppression of desires.
6. Principles: Having a principle such as immortality, intelligence, health, or wisdom.
7. Education: Great knowledge such as intuition, calculus, medicine, and knowledge of infinity.
8. Ideas: Ideas, ideas, ideas!
9. Activity: Meaningful activities such as art, poetry, blogging, and making lists.
10. Greatness: Seeking greatness, no matter whether it's for acting, baking, philosophy, or physics, etc.
11. Role Models: Imitate someone who you think privately failed, but who is a great superficial example, like a gay person who made someone pregnant, or a comedian who was once great.

12. Rewards: Be honest with yourself, and reward yourself when you're honest. For example, a straight man who enjoys self-pleasure or looking at women.
13. Un-Attachment: Focusing on nothing at all, tranquil bliss. For example, not doing math if you hate math. Not socializing if you hate socializing.
14. Focus: Focusing on what matters most can be important. Getting the job done.

FREE WILL HEURISTICS

1. Power has potential.
2. Games define potential.
3. Defining a game defines the potential of potential.
4. Games can be 'left operating' which in the holographic universe as well as the magical universe and the divine universe does not require energy. Also, in the materialist view there would be no overwhelming principle to oppose individual will except limited awareness or physical obstacles.
5. There are at least two forms of will: freedom and preferences.
6. As far as selection within free will or preferences, contingency may be required. Contingency is the ability to resist good or bad potential.
7. Knowing good and bad requires judgment, blessing, or harm. Harm is knowledge of bad potential, blessing is knowledge of good potential, and judgment is knowledge of knowledge.
8. The combination of freedom and preferences is the determination of the will.
9. Perfect preferences are not deter-

mined.
10. Perfect freedom is not determined.
11. The range between both perfect freedom and perfect preferences in relation to perfect freedom OR perfect preferences, and imperfect freedom AND imperfect preferences creates free will.
12. It is also possible to have a standard about the success or failure of preferences and freedoms, creating a gradual realization of freedom based predominantly either on preferences or freedom, but not both.
13. The freedom about freedom is the ability to have both freedom and preferences.
14. There is no standard of freedom without the determined will. However, the determined will is also the ability to have freedom about freedom, to have both freedom and preferences.

PROGRAMMABLE HEURISTICS AND BIG IDEAS

REASONING HEURISTICS

1. Everything contains a multitude of smaller parts, the more so if the smaller parts are imaginary. The more something is real, the more it relies on the unreal. However, something must be general to rely on the specific, and specific to rely on the general. The unreal and the real in the sense of atoms appears to be the only real boundary between general and specific.
2. What is practical is always neutral in some way. It joins forces, or negotiates, or serves as a tool, or organizes between things that are different. There is no form of usefulness which is completely opposite of everything else, unless it is a hidden rule of law which mirrors it, or unless it functions separate from it. Therefore, relevance is a large part of usefulness, and nature and meaning are a large part of the remainder. All of these are forms of practicality which have non-opposing functions.
3. Psychologists must be innocent of the soul in at least one degree. All knowl-

edge is compromised in this kind of way. Every discipline would be absolved of itself if it could no longer create itself. Perfect knowledge is not a domain. Therefore, the only real knowledge within a domain is qualified knowledge which rises above the common meaning of domain. But how is the soul qualified? If there is a conflict, the psychologist will focus on it. The psychologist will focus on anything that is not a solution, because the qualified soul is the same thing as not lacking knowledge in one degree.

4. Philosophers believe all kinds of crazy things, especially if they aren't philosophers.

REFERENCES

Chalmers, David. *The Singularity: A Philosophical Analysis*. Consc.net.

Coppedge, Nathan. "A New Invention: Axiometry". Academia.

-------. "Coherent Data Project". Academia.

------. *The Dimensional Philosopher's Toolkit*. CIP: 2013 - 2015.

------. "The Only System". Academia.

Derrida, Jacques. *Afterword*.

Farlow, Stanley J. *Paradoxes in Mathematics*. Dover Books.

Frege, Gottlob. *On Sense and Reference*. Max Black, trans.

Godel, Kurt. *On Formally Undecidable Propositions of Principia Mathematica And Related Systems*. Martin Hirzel, Trans.

Hájek, Alan. "Philosophical Heuristics and Philosophical Creativity." *The Philosophy of Creativity*. Elliot S. Paul and Scott B. Kaufman, Eds. Oxford U.

Harman, Graham. *The Quadruple Object*. Zero Books.

Honderich, Ted, ed. *Oxford Companion to Philosophy*.

Kant, Immanuel. *Critique of Pure Reason*. J. M. D. Meiklehohn, Trans.

Klement, Kevin C. "Russell's Paradox". IEP.

Peirce, Charles. *Reasoning and the Logic of Things*. Cambridge, Mass.

Popper, Karl R. *Objective Knowledge*. Oxford U.

Rescher, Nicholas. *The Coherence Theory of Truth*. Oxford U.

Roberts, Michael. TADS [Object-Oriented Programming].

Schagaev, Igor. *Evolving Systems*. Academia.

-------, Nibojsa Folic and Nicholas Ionnides. "Multiple Choice Answers Approach: Assessment with Penalty Function for Computer Science and Similar Disciplines". Academia.

Sion, Avi. *Future logic: Categorical and conditional deduction and induction of the Natural, temporal, extensional, and logical modalities*. Avi Sion.

Tarski, Alfred. "The Concept of Truth in Formalized Languages." J. H. Woodger, Trans.

Wallach, Wendell, and Allen, Colin. *Moral Machines*. Oxford U.

BIG IDEAS

Nathan Coppedge

THE BIGGEST IDEAS OF NATHAN COPPEDGE

Greatest:

Theory of working **_Perpetual Motion_** machines. (Practical philosophy).
Formula for the souls of literature (spiritual information)

Objective knowledge (absolute truth)

General solution to all paradoxes (general problem-solving).

Formula for answering questions / ideal format (long-sought wisdom).

Solutions To Long-Standing Problems (methodical genius).

Characteristica Universalis (category language).

Evolution only definitely occurs when by some standard it is complete. Each thing fulfilling its own identity is the maximal evolution of anything. The question is really, can I be fulfilled? Fulfillment is the measurement of reality.—*Notes*

Wizard Logic (magical future history).

Applications:

Premise of **Independently-moving planets using perpetual motion**
Perpetual Motion Vehicles

Perpetual Motion Flying Machines

Apparature Mobile Building Concepts

Self-Recharging Batteries

Basics of **Nano Perpetual Motion**

Premise of **Chemical Perpetual Motion**

Exponential Combined Energy

CatSpur Shoes, (mechanical shoes) make it faster and easier to walk, but are total disasters on stairs and elevators.

Math / Science:

Pursuing the Disintegral
The Polycalculus Concept

- Quality Science — *proposed book title around 2009 to 2013.*

- Protomathematics

- Proto-Physics.

4-D Semantics: found its proper place in this writing: Volatile Theory Construction

PROGRAMMABLE HEURISTICS AND BIG IDEAS

- Philosophy of Science: Premier Science, Abstract Polyps, Top Defenses of Science, Coherent History of Mathematics, Ultimate Critique of Mathematics.

Sun Singularity: <u>Platonic Sight</u>

Human Intelligence:

<u>**How To Think Like Various Geniuses**</u>
- Double Golden Age concept.

Attempt at a so-called <u>**Complete Theory of Human Knowledge**</u> (epistemology).

<u>Synergasm</u> / dynamic type of genius.

<u>High-Minded Colundrum</u>

<u>The Second Zero</u>

- Philosophical Razors — *The Dimensional Philosopher's Toolkit* (2013).

Metaphysical Continuum: <u>Nathan Coppedge's answer to How would organic complexity evolve in higher dimensions?</u>

Metabolic Metaphysics: *my earliest use possibly was in The Dimensional Immortality Toolkit (2015) or here:* <u>How can 'one who emerged' be described with one word?</u>

Metaphysical Variables <u>ON QUALIFIED WRITINGS</u>

Metaphysical Semantics: <u>Metaphysical Semantics: Nathan Coppedge: 9781500520748: Amazon.com: Books</u>

- Metaphysical probability wave (?)

Semantic Psychology *<u>Semantics of Psychology</u>*

Calculus of Psychology: <u>The Certain Evidence</u>

- Exponential Realism: *I believe as a keyword on Academia.*

Divine Anthropology: founded in relation to works such as this detailing human history for gods, to provide context for aspiring immortals: <u>FUTURE PREDICTIONS ON THE PREMISE OF TECHNOCRACY</u>

Ideologies:

Value-Ethic: From a perpetusl motion perspective many things amount to the 'value-ethic' of perpetual motion. (—<u>What do you find when you look at the ethics of technology and the philosophy of engineering together?</u>)
- Modular Society

- Asceticureanism

<u>Secular Religion</u>

- Volitional Economics

PROGRAMMABLE HEURISTICS AND BIG IDEAS

<u>Volatilism</u>

- Dimensional Dollar

Urban Romantic Poetry: <u>POET: Nathan Coppedge - ALL POEMS OF Nathan Coppedge</u>

Miscellaneous Works:

<u>History of Philosophy</u>
- Occult Evidence: Psychic Prediction Techniques, Time-Travel, Telekinesis, Study of Enchantment and Sorcery, Language of Power / Practical occult powers. Original Psychology: Construction of the Sublime Soul, Instructions on the Soul / Explanation of the soul.

- Alchemy / Changing the coin of the age, The Grand Work, the 7 Serpents, The Sword of Mercurius… etc.

- Clever Models: Unlikely Constructions with Building Blocks, Dominoes of Increasing Heights, Simple Difference Engine, Principled Asymmetry Device / Foundational physics of perpetual motion.

- Poetry.

- Writings on Virtue: General Ethics, Abstruse Morality, Value Ethics, Metaphysical Morals, Luxury Platform, Hidden Intellect, Emergent Epiphany, Meaning as Non-

Contradiction, Principles of Value.

Graphics & Design:

Hyper-Cubism / beginning of aesthetic logic (with Francis Picabia, Kazimir Malevich, Georges Braque, Juan Gris, and M.C. Escher).

- Bounded Cartesian Coordinates / typological standard.

- Aesthetic Standards: The Value of Judgment, 4-D Semantics, Arbitrary Mathematics, Relative Absoluteness, Disposable Structures, Apex Computing, Interface Culture, General Systems, Standard of Interpretation, Philosophical Razors.

- Design contributions: Naming of businesses (Eb Lens sports brand, Keys on Kites Tattoos, Evolution Tattoos, Alchemy Club, Bobbi's Grocery, many others) and logos (classic Sprint logo), design of outer body of the Cooper Mini car, design for rest-stop sculpture resembling a modern skyscraper (appeared circa Ohio).

- Conceptual Art: Ideal Texture Palette, Categorically-minded art, Unique representation, Significant variation, Value of complexity, Psychological-emotional meaning, Truth to art, Finesse simpliciter, The road within, Agora mote, Fine fences we build.

Virtual Realism: proper place: A Mad Man's

PROGRAMMABLE HEURISTICS AND BIG IDEAS

Plan of Action

Sublime Methodology — *a general theme*

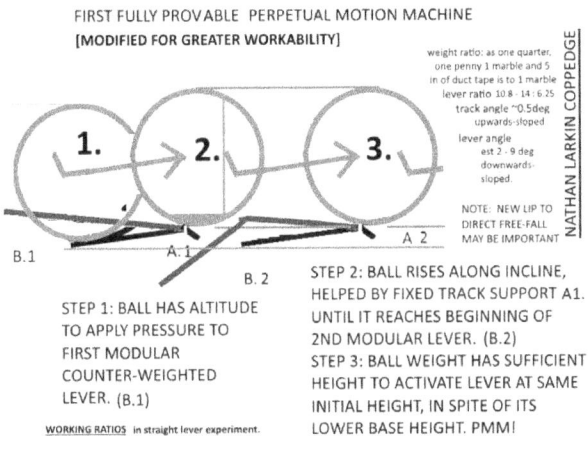

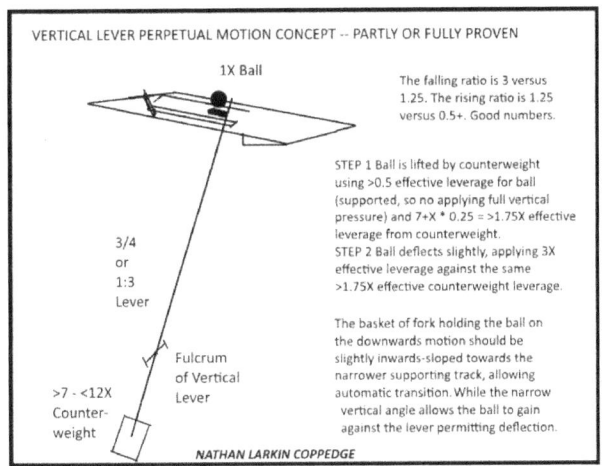

Above: Likely perpetual motion machines (Coppedge)

Nathan Coppedge

Epistemology: <u>What have been the most important works in 21st century formal epistemology?</u>

<u>Premier Writings on Epistemology</u>
<u>How Do We Know?</u>

<u>Definitional Knowledge</u>

<u>The Nature of Meaning</u>

<u>Valuable Notes on the Philosophy</u>

…

"[I can imagine a] Double Golden Age"—
Nathan Coppedge, July 2017

I suppose if a quantitative, abstract infinity, were further abstracted into proportional numbers (which I may have invented as an analogy to Bounded Cartesian Coordinates, which I also invented), then infinity might be viewed geometrically. --<u>Can Infinity be greater than itself?</u>

Thoughts Growing Up (1982 - 2001):

Conditional Nihilism—1983

Sufficion—1983

Coincistance—1983

S'fusion—1983

PROGRAMMABLE HEURISTICS AND BIG IDEAS

Possibility of qualified volition—1984

Direct allegory—1985

Philosophy ==Unique—1985

Fallen from the process—1986

Thought travels like a devil to paradise—1987

Lies are the semantics of math—1988

i = infinite impossibility (1990)

The Chinese invented virtue (1990)

Religion is for immortals (1990)

Pleasure is fake pain (1990)

Magic defines Possibility (1991)

I can be old before I am young (1991)

The soul must be what the soul experiences, even paradoxically (1991)

One thing can be a unity of things outside itself (1992)

Emotional physical immersion (1992)

Laws of nature are games (1995)

Words were once made of problems, but problems are not just words (1996)

Alchemy just requires the color yellow (1996)

Everyone wants to be an enchanter (1997)

Good devil is a devilishly good idea (1998)

Infinities stand for millenniums (1998)

Philosophy is for immortals (by 1999)

If the world survives I can survive (2000)

Perpetual motion is possible (2000)

(—As listed in History of Philosophy by Coppedge).

Added Details:

He's a Time Traveler: <u>Are there any real life incidences of time travel?</u>

1400+ Works of Abstract Hyper-Cubism.

Coined the phrase 'And stare ope-mouthed at those chill jaws' which meets Roethke's unfulfilled quest for an all-accented eight-syllable line.

Inspired brands such as Cooper Mini and Eb Lens, Keys on Kites Tattoos and Evolution Tattoos (I don't like tattoos, maybe you can tell).

PROGRAMMABLE HEURISTICS AND BIG IDEAS

Suggested title for book The Universal Computer by Martin Davis.

Designed a statue that appeared in rest-stops in the Ohio.

Excellent Writings:

Socrates, On Ethics

How To Think Like Amadeus Mozart

How To Think Like Von Neumann

Abstract Polyps

Morality Anticipating the Beiträge

The Deontology of the Will

The Construction of the Sublime Soul

The Correlation of the Sublimated Heavens

Letter Addressed to an Aspiring Wizard

Eastern Sorcery: Siddis

Advanced Magical Rules

Sorcery Part 2: The Wizard's Journey

Sorcery Part 3: Magic Steep

Defining 'Human Nature'

Misc Artistic

Gallery 2: Hyper-Cubism Gallery

Sublimism: Sublimism: Sublimist Art, Architecture, Morality, and Poetry: Nathan Coppedge: 9781508564188: Amazon.com: Books

Aesthetic Logic: Logic School (Facebook)

Pure Formalism: The Art of Abstract Calligraphy

Criticism

Essential Criticism: The Dimensional Philosopher's Toolkit: or, The Essential Criticism; The Dimensional Encyclopedia, First Volume (Re-Issued Paperback Edition): Nathan Coppedge: 9781494992934: Amazon.com: Books

Esential Association: Premier Knowledge Studies 2

Soupcon Aesthetic: perhaps beginning with an artwork from 2005/08 called "soupconcrete".

How to Write Aphorisms: How to Write Aphorisms: The Aphoristic Method; A Guide to Writing Aphorisms: Nathan Coppedge: 9781494799359: Amazon.com: Books

PROGRAMMABLE HEURISTICS AND BIG IDEAS

High-Minded Formalism

Advanced Art

Ideas I have thought about but may not have originated:

What are the theories of sociology and their relevance to the society?

The History of Gravity (predicting future physics)

The Constructor Theory

Edmund Husserl

Socratic Writings

Michael's Calculus

What does identity of indiscernibles say?

Exceptional Morality

Learning Atheism

MISC ALPHABETICAL INDEX (NOT ALL MINE):

2-d, Theory that it is Equal to the 4-d through a hyperbolic flip.

4th Dimension Compatible with Chronology through Chronological Non-Contradiction

Determined in the 4th Dimension: <u>Would it be possible to make an actual, physical simulation of the fourth dimension?</u>

6 Types of Health.

Ages of History, Permanent.

Arbitrariness, Quantum Standard, Against Intermediates, Mereological Modifier (Hard Problem 23)

Arrow of time: Technically it's in the memory or forgotten. The arrow of time is a metaphor for things generally changing. They don't change back essentially because something may have happened. Thus, time travel is the ability to undo causal reality. Time-travelers usually underrate how divine or magically desperate or lucky they are. --<u>When you think about yesterday</u>

Blindspot, Sustenance.

Brain, Implied.

Brain, Universal: From sensitivity formula: sq. rt of the degree of absorbancy.

Bricks, Pouring water over to make easier to cut and reducing noise.

Bureaucratic Language, Universal.

Cause, Effect independent of, separated and

PROGRAMMABLE HEURISTICS AND BIG IDEAS

 thus realized: <u>Incredible Systems</u>

Child, Godchild, Hivechild.

Civilization, Stage of Wizard Development.

Cocaine, Quick-Acting Gradually Sensitivizing, Theory of (in past life as Marie Antoinette).

Complexity, Trivial, in the Year 3000, Imperative for.

Compusition. Formal computation.

Creativity, 'Dearth of'.

Different 'Cool'.

Dimensional Philosophy.

Earth is Analytic, Theory that.

Energized Emptiness.

English, mottled.

Everything, That we were once everything and it was boring theory, that we are now classified as more than everything.

Evolving, in the past.

Extension, No manifestation without.

Females live longer because they have more fun.

Free will, Normative.

Fulfillment, as measurement of reality.

Garden, Forsooth.

Genius Yet Basic (GYB) like perpetual motion and absolute knowledge and M.C. Escher's drawings, Genius Yet Genius (GYG) like large batteries, Basic Yet Basic (BYB) like reusable clothing, Basic Yet Genius (BYG) like labyrinths … Ideas.

Harmonious Guilt (jk rowling).

Historical Potential.

Humility, Pretensive.

'Humorous Occult' Perspective.

Hyperapplications.

Hyper-Matter: Bernard Stiegler's idea: Hyper Materialism (I first really noticed it in April 2019, although I may have had similar ideas as early as 2009: Big Ideas Part 24)

Hypersonic Hearing might be Telepathic. When they start to cry they're telepathic?

PROGRAMMABLE HEURISTICS AND BIG IDEAS

Imagination About Imsgination.

Immortal Justification, Stage typically reached only by philosophers.

Information, Ragged end of. (Hard Problem 23)

Insanity, as Language Problem in Academia. <u>Is it rational for a lone theorist promoting a radical new paradigm to seek peer review?</u>

'Interdependent net of pseudo-indefinite depth'. <u>What do you have in mind when you think about nanosocialism, technology, and society?</u>

Interface Garbage (metaphysical concept).

The Iye, The structure of the universe, <u>Concerning Leonhard Euler</u>

King-gets-the-crown theory of alien planets.

Knowledge Emporium (reading specialized things for advantage).

Lever, Ideal: <u>The Ideal Lever</u>

Lies, General (—<u>Brian Coppedge</u>)

Limits, as critique.

Logic, 21st Century.

Material Memory.

Math Age, Unwelcome Non-Mathematical Ideas, in the Year 1000, Evidence of Curse of the Millenniums.

Meaninglessness, Collapse Into.

Mental equilibrium theory: that mental problems are common at every level, one of several common mediums.

Metaphorical Processing: Is creativity a scientific advance, or just a popular aesthetic?

Momentum, unit of time.

Mortality, theory that it is caused by a sleepiness drug (YY).

Multiple Intelligences, Normativized non-absolute difference.

Nature Gland, organic explanation for metaphysics.

Necessary hypotheticals.

Numbers, Unfinished.

Object of ascent (Blavatsky).

Objects, emaculate (Coppedge).

Paradine, Ghostly (Phenomenon in the 5th -

PROGRAMMABLE HEURISTICS AND BIG IDEAS

7th dimensions in which matter has spirit).

Perpetual Evolution

Persecution, Religious, in the Year 0, Evidence of Curse of the Millenniums.

Philosophy of Engineering (theory of working perpetual motion machines).

Planarity, spectral (that geometric planes may be insubstantial, creating additional equations relating to insubstantiality).

Pleasure, Scaleless (an example of a big idea).

Proneness to Risk = being unlucky.

Pure Processor (computational ideal).

Purposeful Origin, Principle of: Since evolution is a measurement of being evolved, everything is the best example of itself and nothing else. All opportunity is contained within the world and the principle of purposeful origin. (A kind of theory of nihilistic, abstract, entity-based evolution).

Quantum Darwinism, you always get what you don't want in terms of what you do want, or what you want in terms of what you don't want, or things are difficultly fine or random and determined. What survives is an unlikely survivor in a different category. It's not quantum Darwinism you're wrong about if you're right about something. Everything that is true knowledge is hardly ever observed by everyone. Humble knowledge, if knowledge is not humble, has a choice of whatever is humble or whatever is knowledge. It's like if we have a probability of being disappointed, we're not disappointed. Zero and infinity are true with hyper-randomness. When knowledge is not humble, humility guarantees knowledge. When Y is not X, not X guarantees Y (quantum causality). Not caused is like one quantum particle, not-not caused is like two quantum particles, not not-not caused is like three quantum particles.

Quantum Supercession.

Related Surfaces, metaphysical theory of.

Relative Pain, makes pleasure into substance.

Religious people, recent story hypothesis.

Saying Things.

Science and Coffee Addicts, in the Year 2000, Evidence of Curse of the Millenniums.

Sensation Piece, linked to expert web: <u>A Sensation Piece Ostensibly on Alchemy</u>

Skip-dimension (A dimension that is more dimensional than other dimensions, by somehow moving between locations amongst

PROGRAMMABLE HEURISTICS AND BIG IDEAS

dimensions. Also called a mobile dimension).

Spiritual Computers: <u>What's the best idea that hasn't been invented yet?</u>

Stimulation, Meaningful in the Year 4000, Imperative for.

Sufficiency of Sufficient Feelings, Guilt of adequation.

Surveys, Mental, for ex Logues like longevity.

Sympaphylumos.

Temporal distance (time's arrow) may explain black holes through some form of temporal emanation such as dual-directionality.

Transformation, Strange.

Unpacking, Unique. <u>Sergei Dovlatov</u>

Urbanism, Post-, a place too ideal to need drugs.

<u>Vicarious Wish Fulfillment</u>

Winning without Winning.

Wizards, Emotional Development, importance of (*Yes, my sooth*).

Yahwave: A huge tsunami resulting ftom an

asteroid impact.

...

WORDPLAY:

Ambitrary: relative absolutism.

Hyper-matter.

Historical potential.

Material memory.

Inexorable imperative.

Intensive imperative.

Intractible imperative.

Subtle imperative.

So-Called 'Humorous Occult Specialist' (Perspective).

Wall, Second, False.

...

Greatest shower thoughts: modular economics, perpetual motion capitalism (one after the other).

BIG IDEAS

BIG IDEAS PART 1

Practical Metaphysics. For example, karma, and ideas to change the world.

Broken system. A system that tests the limit of reality. An enjoyable system.

Metaphysical work. Work that means something ultimately.

Compositional determinism. Choice between futures / alternate singularities / alternate identities.

Second center. Hierarchies contained in one object or archetype.

Ideal problems. Problems that in some sense aren't problems at all.

Passive motion. The original concept of transformation.

Manifestial computer. Degrees of manifestation are degrees of computing.

Designer logic. Comes after universal logic.

Rhetorical metaphysics.

Atoms are in defense mode.

Greater inference.

Secret combination.

Continuous termination.

Ritual of rejuvenation.

The Infinite Computer concept of physics.

Universal singularity.

Existence karma (if you suffer, you're allowed to be superior. If you're exceptional, the world improves).

Instantialism. The fact that the outside may remain unchanged. Outer factors remain relevant.

Fillamentary Consciousness: sensing of nerves specifically, sometimes with a magical connotation.

Relative relativism = absolutism.

Exponential efficiency. The real meaning of efficiency.

Economic Physics: the theory that the universe justifies physical resources by mental commitments which originate in desires for control and other subtle or material properties, which in turn create the politics of manifestation. This may assume an eternal universe, or just a wide scope of information. One possible rule would be that what is both real and worthless will experience everything insofar as it is nothing. What is everything will tend not to be real, or will tend to have great value, confirming the economic view of physics. The major opposition comes from worthless experiences of everything, or valuable experiences of nothing.

Semantic Energy: energy that is effective even though it is theoretical.

Virtual Realism: the requisite condition of the 4th dimension.

Free Manifestation: a possible means of extending 3-d properties into 4-d or 5-d ones.

Super-normal: a condition in which the center possesses special attributes, or in which the average unit is dynamic and / or high-functioning.

Negative Semantics: magic.

Mass-Energy: energy arising from clever mass interactions, typically in the form of constant potential momentum resulting from structural interrelations and chain-reactions. It never has inherent energy, and depends on methods of cheating to gain the equivalence of force. It can only be explained as interactions over sufficient periods of time, and the amount of time it takes to complete a cycle is indefinite apart from specific constructions, as it is functional at many scales (Energy that is neither gravitational, electric, chemical, magnetic, or atomic, and may be permanent or 'free').

"English knowledge" for example, formulas for knowledge in English sentences. Inspired by sonnets and haiku, these also make use of logical relationships to complete specialized tasks.

Proportional judgement: The view that logical judgements depend on the scale adopted. As a result, any property analogous to visual appearance or form can be used to remark on the logical significance of the overall figure. Where languages are graphical or symbolic and rational, an analogy to other forms of judgment can be found.

Substantive judgement: Where a particular part of a linguistic or symbolic language has greater importance than the other parts, its significance may be expressed in terms of its relation.

Automatic causation: Where some outcomes always have better results in the long-term, the intermediary process may be deemed inconsequential, and thus, an automatic leap to the conclusions is desirable.

Retrofitted ideas: Ideas might be indefinitely improved through a change in representation, or reach a stage of perfection. History involves a dialectic between the best available ideas and the current lengua franca.

Inspired universe. The concept that properties and information evolve through interactions with other properties and information, resulting in a strong tendency for pre-processed behavior. By this rule, problems only emerge from primitive sources, and metaphysical problems are a result of economic physics and comparative evolution.

1-Degree Absolute Knowledge: Categorical deduction.

Method of Paroxysm: A solution to paradoxes. Also, a double-paradox.

Psychic Prediction: Eight interrelated methods may solve much of the problem.

Formula for the Souls of Literature: A method devised for reproducing parts of ancient texts, and maybe defining souls in general.

Intuitive Calculus. How to gain mathematical intuition really fast: <u>Intuitive Calculus</u>

Intuitive Physics. How to gain physical intuition really fast: <u>Intuitive Physics</u>

'Pataphysics. A crazy, perfect form of

knowledge I didn't invent: Nathan Coppedge's answer to What is the best definition of 'pataphysics?

Gestalt Theory. Gestalt Theory (Psycho-Analytic Systems).

The Volit: The 2nd Zero or 2nd Center, a way of quantifying internal progressions. For example, a paradigmatic concept such as Knowledge, Evolutionary stages, Libraries, or Systems may be incremented to show replacement systems. For example, an empty system may be replaced with a knowledgeable one, a primitive lifeform may be replaced with an advanced one, or a system of random titles may be replaced with an organization of the best qualified examples from all areas.

Exponential equality (populism).

Exponential Studies.

Weather Studies.

Old genes vs. youth properties.

BIG IDEAS PART 2

Common Sense: "You may think the potato chips are somehow bad for Buddhists (or some other group), but it turns out this is irrational thinking."

PROGRAMMABLE HEURISTICS AND BIG IDEAS

Identity is absolute, or else (in some way) not related to itself.

Calculus of Psychology: "We must conclude that the intended outcome may be offset by other factors."

Iteration has at least four types.

Number theory has at least four types.

For the types see: <u>Calculus Resources by Nathan Coppedge on Official Nathan Coppedge Blog</u>

All Paradoxes Have Solutions: "Now we can generalize across all cases to determine that where an opposite can be found for every word in the best definition of the problem, then any paradox can be solved."

Dark Matter May be Made of Nothing, or Nature Will Never Be Destroyed: "Either substance is timeless since time is not universal, or time elementally opposes all of the substance of nature. However, we know that time is not a universal form of cause and effect, so if time destroys nature but substance is in some way absolute or infinite, then nature will never absolutely be destroyed."

Only Philosophers are Real in the 3rd Dimension: "Queen ants experience the 2nd dimension, and we too experience the 2nd dimension when we engage in sensation. If Plato's Cave is a singular historical account of the evolution of the only sentient beings into philosophers, then only philosophers are real in the 3rd dimension."

God Invented Evolution: "Almost anything can come into existence through an exceptional law which simply states that it happens to be true by some authoritative definition. Without this rule natural law would be far more natural, and evolution would be a largely unnecessary exception."

See Also: [Knowledge Koans (Enlightened Thought) by Nathan Coppedge on Official Nathan Coppedge Blog](#)

BIG IDEAS PART 3

Change of Focus.

Clear Systems.

Ideas With Their Element.

Otherworldly Symbol.

Modal Operators.

Miscellaneous Attempts.

Journey of Ideas.

Exponential Reminders.

Idees Portentes (Dangling Visions).

BIG IDEAS PART 4

Neural Networking (location by brain).

Exponential Status Transit Code (e.g. info-mering).

Dropped References.

Protege Reality (benefit-for-benefit logic loop).

Prodigy Code (how to be a prodigy).

Decompensating-Compensation (physical or economic or ethical or practical rewards for transactions, motions, initiatives, and sketches).

Telamote: Moving conceptual centers, theory-contingent.

BIG IDEAS PART 5

3-Reactions (dash reaction notation).

Soap is one kind of genius that requires natural genius.

To contain the fundamental purity.

A philosophical property: to carry through after carrying through is understood.

Wisdom like pearls falling.

Disgust with successful intelligence (efficient).

Cycling among worlds.

Thoughts of the world.

That shapes must change dimension to retain form with a different proportion in the same perspective (solution to the Von Neumann Paradox).

BIG IDEAS PART 6

Equal-energy Particle Theory → Singularity → Black Holes.

String theory is a better foundation for relativity than relativity is for string theory.

Quantum is supposed to mean logic.

Quantum logic is better than analogical linguistics: in this way physics is the first to use quantum language.

Quantum language does not return quantum, but relativity, absolute knowledge, string theory, and metaphysical semantics.

String theory involves love of gravity and Platonism.

BIG IDEAS PART 7

Consciousness may come about because of the fact that it is interesting to think about.

All problems should be philosophical.

Life is fundamentally composed of efficiency, information, and magic, which are all forms of information. There is no requirement for energy objectively, as irrationally energy may as easily come from non-energy.

Perpetual motion is when there is momentum from rest at no net loss of altitude.

Physics does not always work unless it includes ideas that are bad by some standard → Perpetual motion might work.

Polycalculus: Physics → Number Theory → Qualified or Unqualified.

Genius may consist fundamentally of Artists, Philosophers, Mathematicians, Physicists, and Inventors.

String-theoretic deduction: X theory → string theory theory → space time theory.

NS* (qualified naive realist system) can have a naive realist standard for free, supporting all of naive set theory.

Where Law is Some, Law = Universal,

where All requires Some, and Law requires Universal.

Infinity: a sandbox, a guppy, a philosopher with an appetite, and all forms interlocking.

With a thought-surface, gamut, archetypes, and the spell prexlicam, infinity can be shrunken down.

BIG IDEAS PART 8

Cleavage and information theory.

Are you masturbating? (Psychologist getting on coat) thought of "you're thinking nothing, well I'm not either" and God is thinking.

Sorry, I wasn't thinking.

Where is the red marker when I need it?

Ively boobly baby.

This is going to be dirty.

Sorry professor, are you covering up something?

Only words can tell.

What's wrong with the usual approach?

I'm an absent-minded professor.

PROGRAMMABLE HEURISTICS AND BIG IDEAS

What happens when we're down in the slump?

We go diving for seagulls!

What's wrong with you today professor?

I've bit off more than I can chew. I feel like I've bit the big one!

What's your answer?

I'm trying to put it together.

We're sorry for you, professor.

Oh, it's just me and my messy office.

We want to break a donut with you, prof.

It isn't my kind of day.

Pictures tell a thousand words, professor.

That picture's always been on my wall.

Why are you so grumpy, professor?

I don't know how to fight it. A field change has come over the campus and I'm too old to tolerate it. It's time to move on.

What aroused you so late at night?

The pickles are pickling.

I don't have words for you!

Menarche e los loanos.

Frequent flyer miles?

I want a different steward.

What's up?

Something's burning.

I'm going to break your nose.

Then we're not going to break even.

It's tempting isn't it, watching a young girl, scantily clad, pose for pictures with an unknown stranger, barely guessing his name, and arousing an enticing entourage?

You couldn't guess her name, either!

BIG IDEAS PART 9

Aesthetic realism.

Divine anthropology.

No-space. Time-travel place.

Metaphysical metabolism.

Communication: language as one word.

Hopeless advancement.

High art / high magic.

PROGRAMMABLE HEURISTICS AND BIG IDEAS

Lattice: formula of formulas.

Functional kudos / electronic greeting cards / points-as-tools: tool-points

Ersatz functions, equally playable.

Identical co-existence.

Calculus of variations.

Dialetheianism.

Esoteric Buddhism.

Prose poems.

Elevated normativity.

Style guidance.

Uileinian Games.

Idea Emporium.

Net of Ephesius.

Time's drawbridge.

Solution by riddles.

Philosophers hitting the wall.

Painting in rituals.

Longevity.

Transcendent phenomenal perspective.

Mind over matter.

God chips in.

When uselessness is broken.

When anger brings realization.

Dawning epiphany.

Instant karma.

'My baby is my religion'.

Art is: An alien, A book, and A potter with broken shards.

Prophecies of somewhere you'll live.

Live relevance.

French kissing.

Lucky circumstances.

Big words.

The going thing.

Appreciating circumstances.

PROGRAMMABLE HEURISTICS AND BIG IDEAS

Deep ethos.

Moment of recognition.

Master trickster.

Conspiratorial tone (secrets).

Time of recognition. Bringing up to speed.

All-out competition. Careful selection of winners.

Free for all. To each his own.

Taking advantages.

'In it for the long haul.' Signs of affection.

Rule makers and rule breakers.

Test of maturity. Test of strength. Test of faith. Test of intelligence.

Familiarity.

Love.

Trials & Tribulations.

Loss.

Cruel jokes.

Riddles of the wise.

Wise women and wise men.

Magical strength.

Staying active. Test of time.

Patience.

BIG IDEAS PART 10

Life tends to be for the winners.

Metaphysical semantics is a good idea.

Immortality is the great quest.

True reality begins with 4-d semantics.

Living to any age is magic.

Reality is a bad angle.

In real life, everything works.

Truth is when even the fake—is fake.

PROGRAMMABLE HEURISTICS AND BIG IDEAS

BIG IDEAS PART 11

Mozart

Iterative blasts

Idea Logic

Brain Chemistry

Souped-up Performance (for example, car engines)

'Jitter' high-tech light displays

Paul Simon

High School

Seminars

Program (Architecture)

State-of-the-art interface

God

Evolution evolves

Societal shift

Icons

The right person at the right time

Reality tech support

The Academy (whatever that means)

Intellectual heroes (Nietzsche, for example)

Philosophical problems

Genius animals

A change for the better

Lonely dots / singletons / outliers / singularities

Advanced Chemistry

Serendipity

BIG IDEAS PART 12

Viral genetics.

Viral air-conditioning.

Viral algorithm.

Viral homology.

Viral homogeny.

Viral asemptote.

Viral phenomenology.

PROGRAMMABLE HEURISTICS AND BIG IDEAS

Viral Cohegeny.

Viral Mesegeny.

Viral Hopogeny.

Viral Hegemony.

Viral Psychogeny.

Viral Misony.

Viral Topography.

Viral Technology.

Viral Monotopy.

Viral Mitophies.

Viral Hetasophies.

Viral Psychosophies.

Viral Hetacosmoses.

Viral Psychohetacosmoses.

Constructor theory.

Binary collusion.

Debt moot echo theory.

Ripped temperature theory.

Antagonistic isomorphisms.

Prismatic oscillation.

Universal fear of worms.

Propeur art: formentation.

Ideosophy pragmenton.

Possibility of isophase.

Probability with disjunction.

Order of meritocracy.

Order pedantic.

Order moot point.

Order fatui.

Order emola.

Order picolo.

Order regento.

Order hephelump.

Order credablee.

Order fifth and last.

Order hungry—clean food.

PROGRAMMABLE HEURISTICS AND BIG IDEAS

Order orders.

Order ors d' ouirves.

Order a bank account.

Order peace out.

Order carrots.

Order grimaces: it does the body good!

A faustian choice of festivals.

Primitavara.

Grist vasa vera hemitypafora.

Primitive calculus jokes.

Beached whale jokes.

Bleeched whale jokes.

Formation of the universe--it's a joke.

Witness a joke.

Play a joke, just try it.

We've tried that again.

That may be history, but it's not funny.

That's straight out of the white house.

Nathan Coppedge

No one knows what's certain Debbi Moore.

What's the shielding on this apocalypse?

Has history gone out the window?

Where's the repair man, I've lost my job!

You might need a few extra computer bricks, the thinking kind!

What's oh-for-nothing on Tuesday? Can you think with me?

Pretty figures aren't of flowers…

Fossilize my evidence, it's criminal!

They say you have a brain in your head, well, I have one in my wallet!

Pictures make perfect recipes if you can just feed them into a machine!

Nothing is ever done until Tuesday…

Make a picture over it, see if the allusion matters…

What is quaking up Maui this morning?

Have a half of cranberry suflee, the Moores are dying.

I wrote it just now. I wrote it this afternoon,

and I can't believe it.

I did it. I solved all problems. All problems are solved. Reality will get better. It's like water, pure fresh water-in-my-hands.

I tell you now my mother: perpetual motion is proven! Because of that! Because of everything in a way, but for the moment—because I looked at that lovely, lovely machine.

BIG IDEAS PART 13

Clearing up clutter.

Artificial Humans: The hypothetical theory that humans are in some way fictional or virtual, often with the implication of the priority of experience and importance of informational knowledge that is not currently known, or not by some. The scope of ignorance is both potentially a problem and potentially a program with artificial humans. It is as if the higher powers want us to make up.the facts.

Hyper-Randomness: The idea that if randomness is random enough, then 'randomness has reason':

Contingent Randomness: the idea that a secondary reality can be hyper-random, sometimes solely in relation to the first reality, otherwise known as scalarized arbitrary relations.

Fundamental Barrier: An early barrier to progress defined by rationality and usually just one assumption, such as the craziness of arbitrariness.

BIG IDEAS PART 14

Constructor Paper: Paper that displays dots that develop into lines depending on how you touch the paper, resulting in highly complex abstract 'constructions'.

Magical Blueprint Paper: Paper that is designed, say, with gray symbols, to be easier to make good symbols, art, and origami with.

Paragonical College: A college that really seems like college regardless of skill level.

Isometric Light (Perduring Halflight): light that seems equally bright and dim no matter how dark it is.

Morph-sculpture: A 3-d or 4-d sculpture or display made to look like a morphing shape, often in the shape of a heart or bladder, with colors, animals, etc appearing briefly.

Motif-Apparition: A symbol that appears at the proper time to make a meaningful indication or to announce a change in process.

Precious Fruits: An ideal concept of what lies beyond immediate reality.

PROGRAMMABLE HEURISTICS AND BIG IDEAS

Hangour (non-sic): A particular, almost criminal feeling of the sublime, that accompanies the most intellectual adventures. For example, playing a game near the tree of knowledge, or time-traveling around a barren tree.

Spell-stroke: A transformative realization that occurs while responding to the challenge or danger of another's magic spell. Or also, magical education.

Fateful Charm: When something works magically because there is nothing else left to do. Sometimes assumed to involve matter manipulation or deadstopping.

Deadstopping: Weeding out undesirable futures as an act of magical will. This may be more like selling them than destroying them.

Perdurant Sorcery: Initially, a particular process involving 1. Excitement, 2. Influence (such as greatness, luck, and fame), 3. Control (such as deadstopping or words of power), 4. High magic (such as advanced incantation, healing, alchemy, or transformation), and 5. Permanents (the study of perdurant magic through repeated practice of spells). Optionally, 6. Immortality, as a platform for magical success.

Now that we can see the world as containing magic, various wondrous things can occur, which may be called the 'charmed life', the 'fantasy', 'the incredible', and the 'wonderful'. Think of this a little bit like the knot in the tree in Peter Pan.

Now there is something called magical opportunity which guarantees a desirable outcome, which occurs. This is beyond the level of mere luck, and shouldn't be confised with traditional fateful fortune.

Now the opportunity is for magical design, or else increasing ideas and efficiency. Since the world's resources are magic, this should make life easier for everyone, realistically. So, realistic magic must come about.

BIG IDEAS PART 15
Valuable Notes on Nathan Coppedge's Philosophy:

Nathan Coppedge's Big Ideas Part 15.

1. Ideal Precious Fruits: An ideal concept of what lies beyond immediate reality.
2. Unto and into yourself: An incidental deepness to existence.
3. Formal logistics: reality according to logical priority.
4. Epiphenomenon: a phenomenal event that changes perception.
5. Analytic Systeme: A collection of alternate justifications about epistemic logic.
6. Logic License: Credit to pursue radical

or equivalent models.
7. Premier Idea: ability to find transcendental meaning.
8. Ornate Perspectivism: the ability to add useful complexity.
9. Modal Metaphysics: the ability to make a mindful map of nature.
10. Wish Structure: the structure of weighted outcomes.
11. Eking the Sublime: the creative act that brings one closer to immortality.
12. Transmagensis: an epiphany that one will do exceptional things beyond the average lifespan.
13. Luxury Platform: Life is pointless without a relation to some form of ideal, whether conscious or unconscious.
14. Purposeful Problem: Life gaurantees value, because purpose is more structurally primitive than existence.
15. At Minima Success / Error: The decision to classify as success or error determines philosophical logistics, but not philosophical reality.
16. Exceptional Solutions: ideal problems imply ideal solutions.
17. Definition of Philosophy: At minima the modality of preferring ideal problems unless ideal solutions exist.

Nathan Coppedge

BIG IDEAS PART 16

Problem complex.

Idea-solving.

Puzzle-constructing.

Growth-monitoring.

Regency calliphate.

Iota obsequay.

Peace-giving.

Ambiguous garden bed.

Soldier's last sojourn.

Difficult sequence.

Savoir faire.

Promised land.

A picture that smells like onions.

Made the crossover.

Folding chair problem.

Belief in delusion (Sea of Ghosts).

Infinite emptiness.

BIG IDEAS PART 17

Also called Transcendental Truth / (~transcendent truth)

[Coherent Systems 2.A.2.C.2.]

…

Alien logic.

Alien polytics

Abstract polyps.

Incoherent exceptions.

Logical Specialism.

Pure Theoretics.

Wise math.

Physical Philosophy.

Metaphysical Variable.

Abstract-Concrete Dimension.

Formula of the Known.

Formula of the Severene.

Formula of the Grotto.

Formula for the Soul.

Itemized Agenda.

Reach Theory.

Idea Theory.

Clausal Relationships.

Typical Turns.

Shapely Truths.

Visionary Quest.

Transcendental Setting.

Divine Signs.

Slow Breathing.

Mundane Formal Relationship.

Scorn of Strangers.

Holy Way.

Divine Clay.

Figurehead.

Godhead.

Eliope.

PROGRAMMABLE HEURISTICS AND BIG IDEAS

Systems Meniscus.

Archetypal Limits.

Enmattered Archetype.

The Transcept of the Known.

The Gnomon of Brilliance.

Outside Architecture.

Raw Idea.

Isopes.

Limited Science (telescoping binoculars).

Essential Space.

Close Logic.

Far Logic.

Intermediate Answers.

Double Logic.

Serene Logic.

Transcendental Logic.

Logic Leap.

Sublime Grotto.

Meta-Gates.

Flight of Thought.

Fanciful Logic.

Construction of the Sublime Soul.

Correlation of the Sublimated Heavens.

Watercolor Machine.

Limned Logic.

Universal Order.

Transcept of the Transcendental.

Divine Leap.

BIG IDEAS PART 18

Abstract technology.

4-d semantics.

Cultural networks.

Mechanical Knowledge.

Ideal Science.

Perfect Science.

PROGRAMMABLE HEURISTICS AND BIG IDEAS

Perfect Education.

Biological metaphysics.

Intellectual shock treatment.

Obscene grammar theory.

Metaphysical Criticism.

Data omniscience.

Direct chemistry.

Autobioticism.

Semibiology.

Neuralism.

Batch ideas.

Starcrap.

Illuminatid.

The Ironing Hand.

Tiedie Living Universal String Theory.

BIG IDEAS PART 19

Idyllea—ideal idea.

Treech—natural teaching.

Idealah—unimpeachable ideals.

Intermediah—unimpeachable media.

Metra—urban meta.

Modia—inter-modal.

BIG IDEAS PART 20

Survival of universal singularities == rare common garbage

All true zero-dimensional constants are infinite.

Strange particles: typical exotics.

Any mere number can be exceeded by raising the reality quotient.

Quantum 3-reactions predict perpetual motion and are not against nuclear chemistry.

Black Shoales: Arbitrary when infinitely divided.

PROGRAMMABLE HEURISTICS AND BIG IDEAS

Cause of depressurization is chemical pressure.

Symmetric readings are useful for 3-d processes, especially with organic shapes.

What is not determined is unknown → Epistemic justification.

Intellectually psychic.

Property-quality.

BIG IDEAS PART 21

Immaterial presence.

Infinite limits.

2-consciousness.

Ethical nobility.

Universal composition.

Physik Princip Derr Philosophy

Shared modes.

Unmasked derivatives.

Ennobled A.I.

Quantum Algebra.

Filligree complex.

"Gods have cold stoves that still cook." — Magical Realism

Magnification vision.

Active transcendence.

BIG IDEAS PART 22

[You're... That's great].

This secret would be one of many kinds of language.

With sufficient resources, every object could be given language and identity. What are their calling signs is a common logic…

The brain could exist as an external perception, an augmentation, computers could be external brains, with consciousness dwelling externally. How to augment the spirit (with the machine)?

The possibility of living between these ideas… emergent processing, silent inflection, logically integrated…

It is like Brahma!

BIG IDEAS PART 23

Animals are phenomenology. Humans are different than other animals.

Genie editing

Architecture gene.

How to Talk in Parables

Secondary:

Survival involves: 1. Knowing danger, 2. Knowing how to cope in general, 3. Making smart progress, and 4. Knowing what's important.

Virtusl objects: the things around them might be real.

BIG IDEAS PART 24

April 1, 2019

Dimensions of X Philosophy.

Hyper-Matter

Material Memory

Perpetual Evolution

Philosophy of Engineering.

Planarity, spectral.

Pure processor.

Skip-dimension.

Temporal distance (time's arrow) may explain black holes through some form of emanation such as dual-directionality.

Pure Research (Research-tap brains).

Independently-powered wandering planets.

Historical Potential.

'Humorous Occult' Perspective.

Treacherous knowledge.

Meaningful particle (2014).

BIG IDEAS PART 25

Eta, how to walk metaphorically on thin air.

Related Surfaces, metaphysical theory of.

Santidotes.

PROGRAMMABLE HEURISTICS AND BIG IDEAS

Compusition formal computation.

Rotor Boat: perpetual motion water vehicle.

Rationaire perpetual motion land vehicle.

Airkite: perpetual motion flying machine.

Interstellar globula dust cloud.

Diffuse Intermedium, space common between multiverses.

Nature Gland, organic explanation for metaphysics.

Child, Godchild, Hivechild.

Logic of the universe is that it is essential causality.

Logic of the mind is that it is a tunnel.

Logic of the world is that it is a triangle.

Logic of location is that it is the 4th dimension.

BIG IDEAS PART 26

I'll try…

- Doilies are images of God (unoriginal, I think my brother or stepmother thought of this).

- The future is the prow of ships near the coast. Kinda unoriginal, fronts of things are already viewed as the future.
- The sun has an antimatter core. Too physical. A philosophy major I knew thought of this.
- Necessity is a group of particles in each person and thing that perfectly express necessity. A bit too physical, and much like Spinoza.
- We share the future, for it is the virtue of discovering the present. I think my advisor thought of this.
- We are a network of souls in a library. The library is the objective structure. I and others thought of this years ago.
- The theory that every person suffers pain to compensate for the pleasures of their parents, and pleasure in proportion to their parents' pain, and this rule follows secretly no matter what physical requirements permit such pleasure and pain. For example, if someone is a real consumer, their child might become a painful semi-conscious black hole. I think my unborn son thought of this.
- The self is a form of art existing for convenience. How about that? Actually, I think my advisor thought of that too.
- There is one master painting that controls all paintings, and paintings are the occult underpinning of reality. I think J.K. Rowling thought of this.
- The nth book creates the rules. I think my brother's friend Larson thought of this.
- Perpetual motion flying machines. I

PROGRAMMABLE HEURISTICS AND BIG IDEAS

> thought of this in detail on May 15 this year.

General formula so far:

'Spatial thing, special part, quality, explain'.

Formula due to my brother Brian from years ago

(reverse it: stories, seekers, special room, to be there: and you get a formula for seeking).

New formula:

'General thing, structure, purpose, experience'.

(New? Probably not).

Idea: universal open conduit, assimilating, desires, love.

Conduit of love? Not very original metaphor.

Painful conduit of impossibility?

Structure of impossibility? Expressed by walls. My advisor explained this. I should have known already, I had written about the Wall of Impossibility.

The mentality of one thing stands out is just about used up.

Pixels? The impossible pixel? My advisor's idea.

Colors that are understood but taken to mean black and white? An understood variation of Mary's Room. They look black and white or not in various ways. Seeing colors is seeing colors.

Colors metaphorical for black and white? I originated this with my advisor years ago, although I'm not sure.

Black book metaphorical for white? Covered by yin yang and photo negatives.

How about a resistive metaphysical strategy thst is designed to select the missing category after some immense amount of time? This is perhaps better subsumed by Douglas Adams' idea of an immortal on an endless to-do list through the universe.

How about a solution to the devil's arithmetic? Thought of this several years ago. Had something to do with randomizing (I think, and then choosing a different number that may or may not be based on it).

How about a basic thing that helps all basic things more than an advanced thing helps any advanced things? I think my brother thought of this years ago.

How about a meaningful machine that gives instructions on how to design itself? This is reminiscent of the Mandelbrot Set and the Peano Series as well as music.

PROGRAMMABLE HEURISTICS AND BIG IDEAS

What about music tone colored buildings? This is covered by the word tones.

How about logical layers of paper? This is covered by the word schematics.

I guess we should start with terms that seem brilliant, and it ends up seeming like 'universal fool' or 'practical jackhammer hat'. Those are covered by cartoons and applying universal to everything.

Sufficient black pixels create technology, what about that? That's similar to a recent philosopher's idea of hyper-substance.

So, let's say, black pixels are the music of the spheres. I think my brother had this idea in high school.

Black pixels are metaphysics? Awful idea, covered by aesthetics and the existence of philosophy videos.

How about, the universe is made of mechanical diagrams, and it doesn't matter whether they're real or not? I thought of this with my advisor years ago. It is similar to Leibniz's *characteristic*. Also, it is basically logic (Aristotle).

Working perpetual motion diagrams make up reality? Well, the "God's pocketwatch analogy" is ancient and similar.

Sending out little lassos to convert things by some unknown process? The religious call it sin. Psychologists call it dependence.

New specifically arbitrary explanation is an unseen logical product of anything other people recently said? I thought of this before and my brother perhaps too parsimoniously called this stupidity or randomness or Chinese room. I rephrased it as an arbitrary product and his mind was blown. But then he said, arbitrary byproduct? Randomness. Random X.

There is a color palette, like cheap watercolors. However, one of the colors is secretly selected among a pile of hidden colors. The rest of the palette is normal. The question is, what is the influence of the hidden colors on the actual painting? What is the logic that can be ascribed to the hidden colors which corresponds to a real logic in the painting? Say, if there are layers of painted lines on one surface, what is the mathematics of the determination that the secret colors play specifically a logical role? Could it be that if we look for mathematical relationships we will never find logical meaning? It turns out this is a traditional advanced question in philosophy of mathematics called the painting problem related to Cantor's diagonal argument.

Music: say you have a different key on the piano for every experience, and each experience seeks harmony, but the music must be composed with variation that might be repetitive. Can harmony be found without repetition in

PROGRAMMABLE HEURISTICS AND BIG IDEAS

the real world, not just a piano? Wouldn't the unique world repeat the music? But it turns out this is a variation of the information paradox.

What about four degrees of substance? But Leibniz thought of infinite degrees in Monadology. Descartes thought of a multitude and Anaximander thought of one infinite substance.

A specific formula? That might be tougher, but scientists call it nonsense.

Ideography? It has been known to be investigated by artists at Harvard or biographers. I suppose it could be the same thing as symbology.

Gravity differential? Just a potential term in physics.

Disintegral? This term I appear to have invented a few years ago, let's invent a variation on it.

How about 'religious disintegral?' There!

Religious Disintegral: Tentative definition: Something absolutely needed for one thing that is absolutely not needed for another.

Nathan Coppedge

BIG IDEAS PART 27

Ancient Greek Theory of Everything.

Hegelian Theory of Everything.

Medical benefits of working perpetual motion.

OU ratings for Coherence or similar.

ARE THERE ANY ZERO-TO-ONE IDEAS THAT EXIST ANYMORE?

THE PERVERSITY OF ZERO?

Concerning mathematics and the actual existence of zero:

The concept of nothing was invented with the zero.

In fact, we could choose to believe that outer space is made of invisible ice if we had the inclination.

Likewise, there is nothing which really says that nothing has to be really small, so some cultures may think nothing is a big number, and that infinite and infinitesimal mean the same thing.

All the more evidence that knowledge should be grounded outside math.

An arbitrary number theory.

EVIDENCE OF PERPETUAL MOTION?

https://www.quora.com/Have-there-been-any-actual-demonstrations-of-over-unity/answer/Nathan-Coppedge

SOLUTION TO THE ABRAHAM-MINKOWSKI CONTROVERSY?

I only have one citation for philosophy of physics, just saying...

I would suggest that the electromagnetic momentum simply involves quantum 3-reactions.

Another conclusion is that the researchers have discovered a fundamental geometric-electric basis for perpetual motion.

In all likelihood the effect will not produce a reactionless drive unless it is designed to be somehow perpetual.

LOGICAL SOLUTION TO THE VON NEUMANN PARADOX?

Von Neumann seems to assume infinite points are divisible by a ratio to infinity.

If we divide the subject figure (say, a unit circle) into four equal pie-slice sections, it is not provable that there are alternate versions of the four sections unless we assume rotation.

As the extension of pie-slices into larger pie-slices seems to imply rotation, some sort of axial modification is logically being assumed in the modification of the pie-slices into the unit figure.

Thus, what seems to be the case is that Von Neumann is assuming that the 2-d figure has a third dimension, or a fourth dimension without a third, or a fifth without a third or fourth, etc.

Ultimately, then, the idea that the paradox is a paradox depends on a false idea of infinity, an infinity in which proportions cannot be taken, and thus, in which rotations can exist in the 2nd dimension without preserving proportion, suggesting greater dimensions.

NEW RELIGION?

I also may have invented the Asceticurean religion ("Ascetic Epicureanism").

Nathan Coppedge

[END OF TEXT]

PROGRAMMABLE HEURISTICS AND BIG IDEAS

RECOMMENDED READING

BOOKS BY NATHAN COPPEDGE

Nathan Coppedge

PROGRAMMABLE HEURISTICS AND BIG IDEAS

Nathan Coppedge

BIO STATEMENT

Nathan Coppedge, b. 1982, is an author and member of the International Honor Society for Philosophers. He is a famous quotable, and visionary artist. His recent project has been to re-construct the lectures of Socrates. He lives near Yale University.